Books are to be returned on or before
the last date below.

2005 DUE

Developments in Rapid Casting

Rapid Prototyping and Manufacturing Association

In January 1997 the two existing rapid prototyping organizations in the UK combined to form the Rapid Prototyping and Manufacturing Association (RPMA) in order to maximize the benefits to UK industry of this rapidly expanding technology.

The new organization focused on promoting and developing the adoption of the rapid prototyping technologies and keeping its members abreast of new developments and system enhancements, while sharing experiences through a comprehensive programme of excellent technical seminars. The membership is administered by the Institute of Mechanical Engineers and membership entitles RPMA corporate members to send up to six members of staff to six seminars a year, or individual members can attend all six events on a individual basis.

The RPMA provides industry with an unbiased forum for the dissemination of the latest technology on rapid prototyping and rapid manufacturing. The most important member benefit is free attendance at a varied and useful seminar programme where members can hear about the most recent developments, and also network with each other in a more informal way.

In addition to the seminar programme the association provides its members with sources of information such as design guides on rapid prototyping and reverse engineering. The website also provides links with other organizations worldwide, sharing chat pages whose purpose is to share news, thoughts, and ideas on rapid product development.

Membership is open to all those interested in the technology and its applications and includes bureaux, vendors, consultancies, academia/research, and end users.

Developments in Rapid Casting

Edited by

G Tromans

**Professional
Engineering
Publishing**

Published by Professional Engineering Publishing,
Bury St Edmunds and London, UK.

First Published 2003

ISBN 1 86058 390 3

A CIP catalogue record for this book is available from the British Library.

Printed and bound in Great Britain by Antony Rowe Limited, Chippenham, Wiltshire.

About the Editor

Graham Tromans

Graham has been involved in Rapid Prototyping since 1990, when he was tasked to help set-up a rapid prototyping facility within Rover Group. He was the first person to be trained in using a 3D Systems SLA500 in the UK in 1991. One of the roles he undertook was to educate engineers within the company, and to help them understand the benefits as to the savings in both time and money to be gained by using these very new technologies. One of his functions was to help drive the application development of these technologies within the Rover Group business.

It quickly became apparent that one of the most effective areas would be rapid casting. Allowing engineers to have castings capable of being used as prototype parts fitted directly to prototype cars. These parts were produced at a fraction of the cost and time of the traditional processes being used at that time. Graham carried out a lot of development work with both engineers who required the parts, and the foundries who produced them. During this time Graham also worked closely with BMW, in the technology transfer of rapid casting techniques into engine development.

Over time these processes became part of the design process, and are used today in a wide range of products, from automobiles to medical applications.

In December 2002, Graham left Land Rover to take up a position as Rapid Manufacturing Consortium Manager at Loughborough University, as part of the Rapid Manufacturing Research Group. He has presented papers on rapid prototyping, rapid casting, rapid tooling and rapid manufacture in USA, Japan, and Europe. He was invited by the Department of Trade and Industry, to be a member of two OSTEMs (Overseas Science and Technology Experts Mission) to both USA and Japan to study rapid prototyping. He is currently Chairman of the Rapid Prototyping and Manufacturing Association in the UK and President of the European Stereolithography Users Association.

Related Titles of Interest

Title	Editor/Author	ISBN
Computer-aided Production Engineering (CAPE 2003)	J A McGeough	1 86058 404 7
Global Vehicle Reliability – Prediction and Optimization Techniques	J E Strutt and P L Hall	1 86058 368 7
IMechE Engineers' Data Book – Second Edition	C Matthews	1 86058 248 6
Rapid and Virtual Prototyping and Applications	C E Bocking, A E W Rennie, and D M Jacobson	1 86058 411 X
Rapid Prototyping and Tooling Research	G Bennett	0 85298 982 2
Rapid Prototyping Casebook	J A McDonald, C J Ryall, and D I Wimpenny	1 86058 076 9
Software Solutions for Rapid Prototyping	I Gibson	1 86058 360 1

For the full range of titles published by Professional Engineering Publishing (publishers to the Institution of Mechanical Engineers) contact:

Marketing Department
Professional Engineering Publishing Limited
Northgate Avenue
Bury St Edmunds
Suffolk
IP32 6BW
UK

Tel: +44 (0)1284 763277
Fax: +44 (0)1284 718692
Email: marketing@pepublishing.com
Website: www.pepublishing.com

Contents

Foreword

Rapid casting has developed from the rapid prototyping technologies, which became available around 1988. There have been many developments in the use of these technologies, and the never-ending developments of hardware, software, and materials have been instrumental in the application development in many industries.

Developments in Rapid Casting has been published based on a seminar held in 2001, organized by the Rapid Prototyping and Manufacturing Association in the UK, and was intended to disseminate facts about how rapid casting is being used, and what research is being carried out at this time.

This volume covers a wide range of rapid casting techniques, starting with an overview of available and proven processes. The worldwide acceptance by industry of these technologies is highlighted in the chapter on Soligen (Chapter 9), a process developed and proven in the USA and being used by a number of companies in Europe. There are a number of case studies presented, showing the wide variety of applications. The book also highlights the innovative research being carried out in the development and applications of these new rapid casting processes.

This book should leave no doubt in the minds of the readers that rapid casting is playing a significant role in the development of prototype castings. The research taking place in rapid tooling will have a far-reaching impact on the way we produce tools for die-casting in the future. Also the ability to produce prototype casting in the intended series production material will mean less tooling modifications, with more testing being carried out at a very early stage of the product design cycle, leading to better 'fit-for-purpose' designs.

Graham Tromans
Manager – Rapid Manufacturing Consortium, Loughborough University, UK

Automotive Engineer Editor's Foreword

As the magazine of the industry *Automotive Engineer* has a worldwide readership that is interested in all things automotive. A magazine, though, only has a finite number of pages and so can only address each topic without going into huge detail. However, there are very many subjects that the readership wants and needs to know more about. It is for this reason that *Automotive Engineer* has teamed up with the successful book-publishing arm of Professional Engineering Publishing with the express intent of producing books on automotive related subjects that are far more in-depth and detailed than any magazine article can hope to achieve.

While each book is a standalone publication, bringing them together under the *Automotive Engineer* umbrella will enable engineers, academics, researchers, students, and many others to refer to an imprint that will build into an indispensable collection of books covering a wide range of automotive subjects. *Developments in Rapid Casting* is the first of these publications.

With time-to-market pressures becoming evermore fierce – especially in the automotive industry – engineers of all disciplines are being put under intense strain to ensure that their products are right first time. There is no time for any subsequent design changes and it must be production-ready from the word go. This means that greater emphasis than ever before is being put on the preparation stages prior to a product being signed off and going into production. Consequently this has led to a spectacular rise in rapid prototyping technologies.

Developments in Rapid Casting is therefore an important book as it brings together all the different processes and examines their respective advantages and disadvantages. Edited by Graham Tromans who managed the rapid prototyping and tooling facility at Land Rover prior to moving to Loughborough University, it will appeal to everyone involved in the development of new product whether they be design, research, or production engineers.

William Kimberley
Editor – Automotive Engineer

1

Overview of Rapid Casting Processes

G Tromans

1.1 Overview

Rapid prototyping technologies have been commercially available for approximately 14 years. These technologies have been used in a wide range of applications. The development of tooling in its various forms has had a significant impact on the way both the systems and the materials have developed. Over the last seven years the use of rapid prototyping technologies in the foundry industry has played an important part in the development of new materials and new systems, that are specifically aimed at casting in its various forms.

Functional metal parts have always been required in the production of prototypes at a very early stage of the design process. Traditionally the metal components were CNC machined if only one was required. This could be a costly process bearing in mind the size of the metal billet that was required and the machine programming. To manufacture several components then wax injection tooling would normally have been manufactured for investment casting, or patterns and core boxes for sand casting.

The ability to write off the cost of this tooling over a limited quantity of parts is very difficult, and in today's tight budget constraints and ever decreasing time scales a new method of manufacturing rapid castings was needed. The following report plans to highlight these new and developing methods of rapid casting; although most of the technologies will be covered it does not claim to be the definitive answer to all casting issues.

1.2 Investment casting

1.2.1 Stereolithography
Quickcast
Since the early days of stereolithography, investment casting foundries have been using prototype patterns built on SLA machines. These early attempts at using the models as sacrificial patterns were very 'hit and miss'. The models were built using a solid build style,

and so the process of burning the patterns out from the ceramic shell caused a considerable amount of thermal expansion, thus causing the shell to crack. This problem was overcome to a small extent by using the 'flask' method of casting. This involves the pouring of the ceramic into a metal flask, which contains the sacrificial pattern. Although this improved the yield rate from the SLA pattern, it was inevitable that large numbers of very expensive patterns would be lost, thus making it a very risky process. Another factor, which made this process very inflexible, was the fact that very few foundries in the UK used this method of casting. So anyone wanting to use this process usually had to send parts to the USA or Canada.

It became obvious that there was a large call for stereolithography patterns to be used in the investment casting process. This would require the development of a new way of using the SLA process. At this time a new resin was being developed by the then Ciba–Giegy, now Vantico. This was an epoxy resin, which had a much lower viscosity. To ensure the patterns built on the SLA machines had a much higher yield rate, the patterns had to be much weaker to allow the burnout to occur more easily and to eliminate the expansion of the pattern. This would mean the development of a new build style, which would create a series of interlinked voids, these voids would allow the resin, which now had a much lower viscosity, to drain away from the part after build completion, thus making a semi-hollow part, which would burn away easily. This new build style, named QuickCast 1.0, in conjunction with the low viscosity resin, would change completely the approach to using patterns manufactured using rapid prototyping technologies in the foundry industry. QuickCast was a major leap in the direction of rapid casting, although through time, this build style, which originally consisted of triangular geometry, went through two more development stages. This involved changing the geometry of the build style, the first one being QuickCast 1.1 which changed the build style from triangular to square, the second one changing the build style from square to an 'offset' hexagon, this was called QuickCast 2.0. The reason being to further weaken the pattern more to allow a still greater yield rate.

Owing to the fact that the patterns were now much weaker, foundries became more confident in using QuickCast. This has caused an upsurge in the foundries that are willing to try this new process, and there are now quite a number of companies that have had considerable success in using QuickCast.

1.2.2   Selective laser sintering
TrueForm™/CastForm™/Polystyrene
There are two manufacturers of sintering systems, these are: DTM based in Austin, Texas, who were purchased in 2001 by 3DSystems Inc, the developers and manufactures of stereolithography systems; and EOS who are based in Planegg, Germany. Both of these manufacturers use a laser scanning process; in this case the laser scans a bed of powder or sand to create the pattern.

The use of investment casting patterns from SLS Sinterstation 2000 was first introduced in 1992. The sintering process was then used to manufacture models made of wax; this was one of the original materials to be released by DTM. The development of this material led users into trying it for the investment casting processes like other materials. At that time it fitted in perfectly with some of the investment casting foundry processes. The main problem with this material was that it was difficult to process and the patterns produced were very brittle. This made handling very difficult and owing to the fact that nearly all parts were made away from the foundry environment, transportation of the parts was very risky. In December 1992

Polycarbonate was released as a material on the Sinterstation 2000, the much improved qualities found in this material opened up new avenues for its use within the foundry industry. Such was the success of this new material overall that the use of wax on the Sinterstations was stopped totally.

In December 1995 a major step forward was made by DTM, a new material totally different from anything else was released. This material was known as TrueForm$^{TM.}$ The powder, which was a co-polymer containing Polytsyrene (PS) and Polymethyl Methacrylate (PMMA), gave exceptional surface finish, crispness of detail and better accuracy, owing to its small particle size and spherically shaped particles. TrueForm^{TM} also allowed much smaller wall sections to be cast, owing to the thermal expansion of the material being almost nil.

The greatly improved surface finish of the patterns also reduced the need for surface finishing of the casting, as no support structures are required the post finishing of the models is almost totally removed.

This material has been successfully used by a number of companies, covering a wide range of applications.

In March 1999 a new material had its press release, CastFormPS. This has had very successful trials with wall thickness as low as 1.14 mm being cast. The material, when compared to other casting materials, other than wax, has the lowest ash content after burnout.

One of the main benefits of using this material is the casting manufacturing time. The other processes that require the patterns to be removed in a flash fire furnace, such as QuickCast^{TM} and TrueForm^{TM}, take on the average of 2/3 weeks to cast after the receipt of the RP pattern, this is due to the requirement of a totally dry ceramic shell before the burnout procedure, to stop any excessive moisture boiling within the shell and expanding, thus causing the shell to crack. As CastFormPS is primarily Polystyrene the casting can be produced using the plaster casting process, this allows much lower temperatures to be used in the pattern removal, alleviating the problem of boiling. This can mean the turn around time from receipt of pattern can be less than five days, and the cost is not so high.

EOS introduced their EOS-P system in 1994; this also has the capability of producing parts in more than one material. The material used to support the investment casting process in this case is another form of Polystyrene. This material, like the one previously reported, can also be used in a standard ceramic shell type investment casting or the plaster casting process, owing to its much lower burnout temperatures. One advantage of the EOS-P system is the recent introduction of the EOS P700 this has the biggest part-building capacity to date at the time of writing this report. This allows full size manifolds, and in some cases cylinder heads, to be cast for the automotive industry.

1.2.3   Fused deposition modelling

Patterns created by fused deposition modelling a process developed and manufactured by Stratasys Inc, are also used in the investment casting process. Primarily a wax material was used, and still is to some extent, to manufacture patterns direct from the machine, these could be used in the standard foundry process. The big draw back of using the wax patterns is that of transportation. The very nature of the wax material made the patterns very brittle and easily breakable. The use of the wax patterns in the traditional process meant that the patterns were manufactured from tools within the foundry facility, this minimized the handling requirements and removed any need for transportation across sites.

This problem was overcome to some extent by the development of the use of ABS, another material used by the FDM process, in the production of investment casting patterns. The use of ABS has alleviated the transportation problems owing to its higher strength. Other benefits gained from the higher strength, are the ability to produce much thinner solid walls, in comparison to the resin-based systems, (again due to the much lower expansion rate of ABS), and the parts can be cleaned up more easily owing to the better handling capabilities. The parts also have a much higher surface definition than those produced in wax, owing to the good powdering characteristics of the ABS, which allows final surface finishing to be carried out. However the surface finish is not as good directly from the machine as that of a wax pattern produced in the traditional wax tooling process. The parts can also be built with a honeycomb fill to minimize the amount of material used and to assist with the subsequent ABS removal from the investment shell.

The ABS models produced in this way have to be burned from the ceramic shell in much the same way as most of the other patterns produced by the rapid prototyping systems, that is by use of a furnace instead of using an autoclave, due to the relatively low temperatures used in the autoclave process, and the high melt temperature of the ABS. Again one of the prime factors in the successful use of this technology is the choice of foundry. Not all foundries have experience using this type of material, unlike wax, so until experience has been gained over a period of time, it would be better to choose experienced foundries to ensure maximum yield.

1.2.4   Laminated object manufacturing

This paper-based process also uses a laser to manufacture patterns, the laser cuts successively built up layers of paper, which are bonded to the previously cut layer. This process is particularly good for the manufacture of the traditional type of pattern, which would have been manufactured in wood for the sand casting process. The prime advantage is the acceptance of the patterns within the patternmaking industry owing to the composition of the material used. Traditional pattern makers preferred using LOM models owing to the patterns being easily bench finished.

These patterns can also be used in the investment casting process, although some foundries are not happy with the ash content left behind in the shell.

A considerable amount of development has gone on in this area, with low ash content materials being developed specifically for the LOM process.

1.2.5   Direct shell production casting (DSPC)

This three-dimensional printing process was invented and developed at The Massachusetts Institute of Technology and has been licensed to Soligen based in Northridge California for metal casting.

Where this process differs from the rest is that there are no actual patterns made for removing the shell. This process manufactures ceramic shells that have integral cores direct from the CAD data; these shells are similar to those created by the dipping method.

The system works by a type of printhead moving over a layer of fine Alumina powder, depositing tiny drops of colloidal silica onto the powder in a pattern the same as that of the section of the part. The next layer of powder is applied and the process is repeated until the shell is complete. When the shell is completed the loose powder is removed and the shell fired and poured with metal.

This process is particularly good for complex geometries that need a fast turnaround. Soligen do not actually market this system (at the time of writing), but run a one-stop shop for metal components, offering a complete foundry service also. This can be thought of as an advantage owing to the other systems manufacturers needing to train foundries to work with their processes.

1.2.6   Wax pattern tooling

The use of wax patterns for investment casting dates back beyond the Egyptian Pyramids. In fact it has been traced back to the Shang Dynasty in China, over 3000 years ago. These wax models were formed by sculpting the wax into shapes and then coating with plaster, each one being unique.

Today we need exact replicas, usually produced in high volumes. To this end modern day waxes are produced through the use of tools, usually made from aluminium, but with ever increasing usage of other material such as tooling board, silicone rubber, and many others. Rapid prototyping systems are also used for producing tooling in its many forms for wax pattern manufacture in the form of direct aim (SLA), rapid tool, etc. These wax patterns form the basis of all investment casting processes. It was from these patterns that investment casting has developed into its present day form, using ceramic shells to coat the patterns before being autoclaved to remove the wax pattern. Castings made from these tools will probably have the best surface finish of all the techniques owing to the tool surface finish.

1.2.7   Vacuum metal casting system

This system allows the casting of a variety of non-ferrous metals (excluding Magnesium), using a fully automated, controlled process, which replicates the production properties of aluminium die cast alloys. The process is a complete turnkey operation with parts produced in around three days.

The system uses the same process as a standard foundry investment casting process, using either wax or other patterns that are removable from the ceramic shell. The current machine, the MPA300, holds three litres of molten metal, and the cylindrical flask size is 13 inches by 19 inches in height. This flask is the container for the ceramic shell which is poured around the pattern under vacuum, this ensures an air-bubble free shell, resulting in superior surface finish, giving precise detail in the final casting. The flask is then fired, at which time the pattern melts and the shell cures. The induction chamber at the top of the machine then melts the metal and at the same time the casting flask automatically seals against the upper unit. Once the metal has reached it's required temperature, the plunger rises and molten metal pours into the mould under vacuum giving exceptional casting detail. The total casting time is around three to six minutes.

The flask is then removed from the machine and left to cool to room temperature. Once room temperature has been achieved, the shell is removed from the casting using a high-pressure water jet.

This process allows companies to manufacture high quality metal castings with very little foundry experience.

1.3   Other casting processes

1.3.1   Concept modellers
New low cost rapid prototyping systems are now on the market that have primarily been aimed at the office modeller market. These systems, by the very nature of the materials they use, have opened up a new concept in investment casting pattern manufacture.

Thermojet
This system, developed and manufactured by 3Dsystems, uses a print jet technology depositing a wax material onto a base, building up the part in layers. The down side to this process is that the supports that are built with the part are difficult to remove and require much care in achieving a good quality surface that is to be reproduced on the casting. This part is then used in the investment casting processes; any standard investment process can be used, as this material is capable of being autoclaved. This process has been accepted widely within the industry owing to the ease of use within the foundry.

Modelmaker II
The SolidScape process (formerly Sanders Prototype) uses a thermoplastic material to build the pattern and a wax material to build the support structure; unlike the previous process the support material is different to the pattern material. The support material is removed from the part by immersion in a bath of solvent; this ensures a very good surface finish. The down side of the machine is its build speed, although the high accuracy and surface finish more than make up for this if the technology is applied correctly into the right industry. The jewellery industry is a prime user for this type of machine as is any industry that manufactures small highly accurate castings that require a good surface finish.

Dimension
This system is developed by Stratasys Inc, working on the same fused deposition modelling process as the companies larger systems. The process produces parts that can be used as casting patterns in the ABS material. This material is the same as that used in the larger systems and is not accepted as freely as some of the other materials owing to the special requirements of the foundry to remove the sacrificial pattern.

1.4   Sand casting

1.4.1   Sand sintering (direct croning process)
This process, which has been developed by both of the major sintering system manufacturers, uses a laser to sinter polymer-coated sand for casting patterns. This system actually produces the sandboxes and cores directly from the CAD data, instead of making patterns from which pattern equipment such as core boxes are moulded.

One of the advantages of this process is the removal of the need for geometrical change to the original design by the addition of draft angles, etc. The only areas that need to be adjusted are where core prints are needed. These are added to the original model, and manufactured with the pattern. The system uses both 'Zircon' and Silca' sand, and so can be used to cast both ferrous and non-ferrous material. Again, although this is a fairly standard process to most sand casting foundries, some experience is needed owing to the out-gassing problems. This method is a very quick way of producing single castings in a very short space of time.

1.4.2   Traditional patterns

Traditional pattern manufacture can also benefit from the use of rapid prototyping technologies. The addition of complicated geometrical sections of patterns, that have been manufactured on any one of the rapid prototyping systems, can very quickly speed up the lead-time for pattern manufacture.

Patterns can also be built in total on the systems and then taken to the traditional pattern shop for final finishing, laminated object manufacturing models are particularly good for this application, owing to the wood like finish of the models. These models lend themselves to the traditional patternmakers processes and tools better than any of the other systems.

1.4.3   Plaster casting

The use of the plaster casting process can also be used to manufacture parts, using patterns produced on the rapid prototyping systems as either master patterns or in some cases sacrificial patterns.

1.4.4   Spin casting

As with plaster casting the spin casting process can also use patterns produced on rapid prototyping systems to manufacture castings. In this case the requirements are usually for small detailed parts, which can be produced in Zinc alloys. The process uses the RP patterns to create a vulcanized rubber cavity. This cavity is then spun in a centrifuge and the alloy poured into a central feed. The resulting parts give excellent detail representation and a representative surface finish of the master pattern.

1.4.5   V process

The V process uses the geometry of the actual tools manufactured on rapid prototyping systems to produce castings in full production quantities (30/40 000 castings). The casting process consists of a cope and drag tooling configuration. The cope and the drag are manufactured as an RP model of the tool and set onto a baseboard. This has a very thin film pulled over the tool, in a system much like you would see in the vacuum forming process. A metal frame is then place over the board, while the film is still under vacuum. This frame is then filled with sand, which is compacted down. A film is then placed over the sand and a vacuum pulled through the sand. The frame, complete with sand, is then lifted clear of the tool and inverted. The same process is followed for the other half of the tool, and the final sand, complete with its frame, is placed on top of the initial frame, during these operations both frames maintain the vacuum. The metal is then poured into the sand mould and left to cool. Once cooling has taken place the vacuum is released and the sand falls away from the casting. Castings manufactured using this process tend to display a superior surface finish, owing to the film over the sand face on the initial pour.

The RP tools used in this process tend to maintain a high surface quality with very little wear owing to the protection that the film affords to them during the sand filling and tool removal process.

Conclusion

This Chapter has covered a variety of casting processes that can be employed using patterns produced from rapid prototyping systems. It is clear that there are usually a number of solutions to solve any one casting problem. The choice is then down to two points. First, what is the most easily accessible process? And second, what are the cost implications? although not always in that order.

The overall availability of casting patterns, for whatever foundry process, made using these RP techniques opens a whole new dimension and meaning to the words rapid casting. But let's not forget, unless the foundry industry is willing to be an active partner in the development and use of these types of pattern, the full potential will not develop. The primary drivers for developing this requirement are the rapid prototyping users, industries that require rapid casting, and bureaux that want to add the supply of rapid castings to their list of services. These users, including the system manufacturers/developers, must be willing to work with the foundries and share some of the development costs; either financially or by supplying some 'in kind' contributions.

This Chapter can only be accurate at the time of writing, with developments in rapid prototyping technologies and rapid casting processes moving at the rate they have shown to move over the last 14 years, new processes will be available to add to this list in a very short time.

Glossary of terms

(1) **Stereolithography (SLA)** – This technique uses a laser to scan a vat of liquid photopolymer resin. The resin is cured in layers, each layer being bonded to the previous layer by overcuring.
(2) **Selective laser sintering (SLS)** – This technique uses a laser to bond a polymer-coated powder together in layers.
(3) **Laminated object manufacturing (LOM)** – This technique uses paper layers each one being bonded to the previous one by adding heat. The process then uses a laser to cut the bonded layers.
(4) **Fused deposition modelling (FDM)** – This technique uses a filament of polymer, which is passed through a heated nozzle. Each layer is then built on top of the previous layer.
(5) **Direct shell production casting (DSPC)** – This technique uses ceramic powder bonded together using a binder that is applied through a print head.
(6) **Investment casting** – The process of using a sacrificial pattern, which is coated with a ceramic shell. The model is then removed by using an autoclave or furnace. Metal is then poured into the remaining cavity.
(7) **Sand casting** – This process uses a pattern, which is used as a master pattern to mould foundry sand around, the master is removed and metal poured into the remaining cavity.
(8) **Plaster casting** – This process is similar to that of investment casting, only instead of a ceramic shell being used, plaster is used as the shelling media. The plaster is faster drying than the ceramic shell so giving a faster turnaround time; the accuracy is also usually better. It is considered better for heavy section castings than investment casting.

(9) Spin casting – This process uses a master pattern to create a mould of vulcanized rubber. This is then placed in a centrifuge and spun. The metal is poured into the mould from the centre; centrifugal force then distributes the metal into the cavity.

(10) V process – This process uses a 'cope' and 'drag' type tool which has a thin membrane vacuum formed over the surface. Each part of the tool is then covered in sand, to which more vacuum is applied. The tools then come together and the cavities are filled with molten metal. The vacuum is then released and the sand removed.

Acknowledgement

This 'Rapid Casting – State of the Art Report' was first prepared by the author as part of a European Thematic Networks Project called, **RA**pid **P**rototyping and **T**ooling **I**ndustrial **A**pplications. Ref No BRRT CT-98-5097.

G Tromans
Rapid Manufacturing Consortium, Wolfson School of Mechanical and Manufacturing Engineering, Loughborough University, UK

2

A Review of Potential Rapid Tooling Techniques for Magnesium Die-Casting Applications

R Hague

2.1   Introduction

This Chapter is concerned with potential rapid tooling (RT) techniques that may be suitable for magnesium die-casting applications. This Chapter should be read in conjunction with Chapter 3 that practically assess each of the RT techniques listed in a detailed case study performed for Ericsson Mobile Communications AB[1] **(1)**.

Although not meant as an exhaustive rapid tooling review, the following sections aim to give a brief overview of the rapid tooling technologies available that are considered relevant for die-casting applications. It should be noted that the RT techniques detailed in this Chapter are not *confirmed* as being suitable for magnesium die-casting, but are considered *potentially* applicable by the author, based on the ability to be used in a commercial hot-chamber die-casting machine with standard die-casting magnesium. Many available RT techniques have been omitted due to considerations, such as time to manufacture the inserts and their perceived inability to withstand the harsh environments of a die-casting foundry. So called 'soft tooling' techniques are not considered

2.2   Definition of rapid tooling

Many in industry are now familiar with the idea of rapid prototyping, tooling, and manufacturing (RPTM) and many gains have been made in these fields. At present, many companies are concentrating on reducing the costs and lead times for tooling and are, therefore, more focused on the rapid tooling part of RPTM.

---

1 Note: Ericsson Mobile Communications is now part of Sony Ericsson

The use of rapid prototyping (RP) machines to produce tools is now common and well documented. However, the term 'rapid tooling' (RT) is often used to describe processes that are not produced using additive techniques. Rapid tooling is generally used to describe any process that reduces the long lead times that were traditionally achieved using conventional machining techniques before RP processes were in existence. In this vein, it can also be argued that high-speed machining (HSM) should be included within the RT group as the time gains that have been achieved in the past few years, compared to traditional machining techniques, have been remarkable. However, the subject of HSM is generally beyond the scope of this review, except for the case where HSM is performed using STL files from which to develop the necessary tool paths.

As well as the direct additive (RP type) or direct subtractive (HSM) techniques, it is possible to produce RT mould or die cavities by casting a material around a master pattern (that is very often produced on an RP machine). Tools made with this approach are considered to be 'indirect' RT solutions. Therefore both additive and subtractive processes, as well as direct and indirect methods, are included within this Chapter and are broken down into three main categories.

- Subtractive rapid tooling.
- Direct 'RP' type rapid tooling (additive/layer-wise).
- Indirect rapid tooling.

2.3   Potential rapid tooling techniques

2.3.1   Subtractive RT solutions
2.3.1.1   High speed machining of STL files
This is a technique that is being pioneered by, among others, TNO in the Netherlands. The approach that is taken is to use the relatively 'simple' STL file format as the basis for tool cutter paths for a HSM machine. The use of the STL file, coupled with other techniques, allows a more automated approach to the complex area of HSM programming and processing, thus enabling a greater uptake of HSM by a wider audience, who do not necessarily have the machining background required for standard programming and machining **(2)**. This approach has recently been commercially adapted **(3)**.

2.3.1.2   Laser caving
The laser caving process is a direct tooling technology that uses a 750 W CO_2 laser to ablate (chip-away) the surface of the material being processed to form a cavity. LCTec GmbH invented the process, but it has now been bought and is being further developed by DMG in Germany **(4)**.

The laser caving process is most suited to tool areas that require high accuracy, good feature definition, and good surface finish. It is reported that, potentially, the system could produce dimensions to ±0.03 mm and with a surface finish of Ra = 1.5 µm **(5)**. At present it is only possible to process ceramics and ferrous metals with laser caving as non-ferrous metals tend to reflect and damage the laser system. A view of the type of detail that can be produced by laser caving is shown in Fig. 2.1.

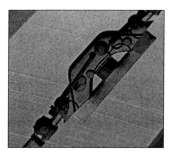

Fig. 2.1 Type of detail achievable with laser caving (4)

However, it should be noted that this is still a developmental type process (although on sale) and there are some difficulties still to overcome. The main problem with laser caving is that, by itself, the process is relatively slow – therefore, the DMG version of the machine incorporates a milling action to remove the majority of the material, while the laser is used to create the fine-details. However, there is a conflict in the current approach as the milling part uses standard cutter paths from the CAD model, whereas the laser caving part uses STL input. Therefore, it is difficult to reference these two packages (note: one aspect that is being pursued involves the HSM of STL files, as listed above, thus making a common standard of file manipulation for programming).

2.3.2   Direct (additive) rapid tooling processes
2.3.2.1   *Laser sintering*
Laser sintering can be generically described as an additive rapid prototyping process, that builds parts (or tools) by slicing a STL file into a series of discreet cross-sections, and then sinters (or melts) a powder layer with the profile of that cross-section. This is done, layer-by-layer until the part (or tool) is completed. To achieve metal objects, there are two alternative commercial types of laser sintering.

• The sintering of polymer coated metal powders, which then need to be post-sintered and infiltrated with copper [3D Systems process **(6)**].
• The direct sintering of a steel powder matrix within the laser sintering machine [EOS process **(7)**].

2.3.2.1.1   3D Systems Laserform ST100
The 3D Systems (formerly DTM Corporation) approach to laser sintering involves using one machine to process a variety of different powders. The machine can be used for processing plastic, sand, and metal materials – however, in the case of the sand and metal materials, they are 'polymer coated' to allow them to be processed in a machine that was ostensibly developed to process plastic materials. The 3D Systems process is known as laser sintering (LS) – formerly selective laser sintering (SLS).

For the metal powders, the polymer coated metal material was originally known as RapidSteel and is generally used for injection moulding tool inserts or for the production of parts. There have been several releases of the RapidSteel material – the latest being known as 'LaserForm

ST100'. It is necessary to perform a series of post-sintering operations to make the 'green' part produced on the LS machine into something more functional. This is achieved by putting the green Laserform ST100 tool/part through a furnace stage to remove the remainder of the polymer binder and to promote sintering between the metal powder particles. After the furnace stage, the part is then infiltrated with a bronze or copper material to make it fully dense. It is necessary to be careful during these post-processing stages to avoid any distortion of the part (or tool).

The advantages of this process are:

• relatively quick;
• reasonable tool strength;
• conformal cooling is possible;
• can be used for different processes (injection moulding, die-casting).

The disadvantages are:

• distortion;
• expensive equipment needed for essential post-processing of the inserts;
• size limitations.

2.3.2.1.2   EOS direct metal laser sintering (DMLS)

The EOS direct metal laser sintering (DMLS) process is an additive 'rapid prototyping' technique that has been adapted for the production of mainly tool inserts. The DMLS process uses a higher power laser (approximately 200 W) to fuse together a metal powder mix, thus producing the metal object directly – without the need for a post sintering and infiltration step. It is difficult, at present, to produce complex part geometries as the DMLS process (unlike the 3D Systems process) requires the use of support structures for over-hanging areas – these over-hangs are not generally found in tools and, thus, the process is mainly used as a tooling system.

Initially, the DMLS process used a bronze powder mix (known as 'DirectMetal') that was useful for short runs of plastic injection moulded products. A steel powder (known as 'DirectSteel') was introduced that builds parts in layers of 50 microns – this material was meant to be more robust than the bronze material and was intended for extended tool life. However, the surface finish is relatively poor and requires fairly extensive hand-finishing to make it useable. More recently a powder has been introduced that allows parts to be built in layers of just 20 microns – for plastic injection moulding, studies have shown that minimal finishing is required of these tools (light shot-peening). However, for die-casting it is still necessary to perform a reasonably high degree of hand finishing.

The advantages of this process are:

• bronze powder is relatively quick – though not suitable for die-casting;
• reasonably good feature definition;
• conformal cooling is possible;
• can be used for different processes (injection moulding, die-casting).

The disadvantages are:

- surface finish of steel powder needs improving;
- steel powder is quite slow at present.

2.3.2.2   ProMetal

The ProMetal process is based on the three-dimensional printing (3DP) process developed at MIT **(8)**. 3DP is an RP type additive process that uses a combination of powder and ink-jetting technologies to produce a part (or tool). This involves laying down a layer of powder and then binding it together to form the finished part.

For ProMetal, the powder that is deposited is tool steel and, after the layers have been jetted together, the green component is put through a furnace stage to infiltrate the matrix with a bronze material. Although the surface finish of the resultant inserts or parts is quite poor, the ProMetal process has the advantage that cooling channels can be placed wherever they are needed (conformal cooling), thus enabling reduced cycle-time due to more efficient heat control.

The advantages of this process are:

- conformal cooling channels are possible.

The disadvantages are:

- post processing is necessary;
- size limitations;
- not really considered as 'rapid' – more aimed at production market.

2.3.2.3   LENS (laser engineering net shape)

The LENS process was initially developed at Sandia National Laboratories in the USA **(9)**. It is effectively a laser cladding system where a high-power laser is focused on to a substrate to create a molten pool on the substrate surface. Metal powder is then fed into the melt pool created by the laser to add material and therefore, increase the size of the component. The part (or tool) is built up by scanning the laser and the powder-feed arrangement in the profile of the part (or tool) that is required – this is done in a layer-wise fashion (as with other direct rapid prototyping techniques) until the part, as represented by the CAD model, is produced.

There are two companies that have commercialized LENS-type processes, both in the USA. These are:

- Optomec;
- POM.

2.3.2.3.1   LENS Optomec

The LENS process in the Optomec form uses a 750 W Nd:YAG laser. The general arrangement and method of processing is shown in Fig. 2.2.

Fig. 2.2 LENS (Optomec) process (10)

2.3.2.3.2   LENS POM (DMD)

The POM process is very similar to the Optomec process, except that a 750 W CO_2 laser is used instead of a ND:YAG laser. In the POM form, it is known as direct metal deposition (DMD). The POM system is shown in Fig. 2.3.

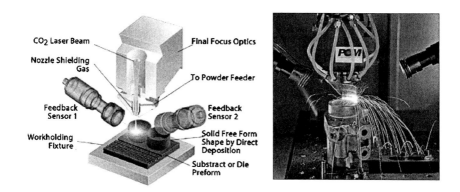

Fig. 2.3 DMD POM process (11)

2.3.2.3.3   Conclusion on the LENS processes

The 'advantage' of the LENS process is that, in theory, it is able to produce parts built in multiple materials, thus giving rise to exotic graded structures that have not been possible before. Also, conformal cooling channels should also be possible.

However, the disadvantages far outweigh the advantages at present – especially for rapid tooling applications. It is currently very difficult to represent the potential graded structures in a CAD environment, therefore, nullifying one of the advantages. Also, as they are near net shape processes, the surface finish that is achievable is very poor and some kind of post machining is necessary – therefore, in most cases, it would make more sense to high-speed machine the parts directly. At present, these processes are probably most suited to repairing existing tools that have been damaged.

2.3.2.4   *Controlled metal build-up (CMB)*

Controlled metal build-up initially operates in a similar method to the LENS type processes, except that, instead of a powder feed into a laser, there is a wire feed (like MIG welding) into the laser. However, unlike the LENS processes, after each layer has been processed, there is a machining operation that smoothes the layer that has just been deposited in preparation for the next layer.

This system was initially developed by Fraunhöfer IPT in Germany and has been commercialized by Röders **(12)**. However, the process is still in the early stage of development. The CMB system is shown in Fig. 2.4.

Fig. 2.4 CMB Process (13)

2.3.2.5   *ARCAM*

The ARCAM process is another metal powder processing method that uses an electron beam, as opposed to a laser, to build up layers of the chosen metal powder. This is a little known process that is still in the development stages – problems encountered include poor surface finish. The process is shown in Fig. 2.5.

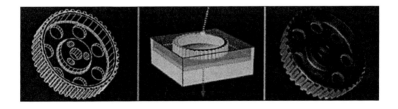

Fig. 2.5 Schematic of ARCAM process (14)

2.3.3   Indirect rapid tooling

2.3.3.1   Keltool

Keltools are made in bronze and stellite. As it is a proprietary process, there is little information available about how tools are made but it appears to be similar to a process described by Ruder (15), as follows.

A powder/binder mixture is poured around a master, with a solvent probably being used to keep the mixture fluid. Once the solvent has evaporated, the binder holds the part as a green compact, in a similar way to that used in powder metallurgy. The part is then fired at a high temperature to burn off the binder and sinter the powder particles together. Copper is then infiltrated into the porous structure. The process is shown in Fig. 2.6.

The 3D Keltool Process

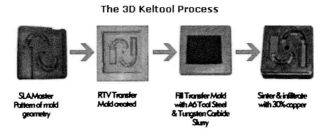

SLA Master   RTV Transfer   Fill Transfer Mold   Sinter & infiltrate
Pattern of mold   Mold created   with A6 Tool Steel   with 30% copper
geometry   & Tungsten Carbide
                                        Slurry

Fig. 2.6 The Keltool Process (6)

The tools from this process show good definition and surface finish. The disadvantages are that shrinkage has to be allowed for in the sintering process, the inserts are prone to distortion, and they are limited to relatively small tools (due to problems of holding dimensional stability during the furnace/infiltration stages).

The advantages of this process are:

•   excellent feature definition;
•   suitable for fine detailed part;
•   short lead time (quoted).

The disadvantages are:

•   size of inserts – limited to about 150 mm × 150 mm;
•   control of the process is difficult;
•   distortion;
•   mismatch between quoted and actual delivery times.

2.3.4   WIBAtool

The WIBA process is very similar to the Keltool process, but uses a different material and is less established. It is being developed by a Swedish company (called WIBA) which is performing on-going research (16).

References

(1) www.lboro.ac.uk/departments/mm/research/rapid-manufacturing/consortium/case_studies.html November 2002
(2) www.ind.tno.nl/prototyping/rapid_tooling/3dcadcam.html November 2002
(3) www.delcam.com/
(4) www.deckelmaho.de/
(5) www.3trpd.co.uk/technical-information.htm November 2002
(6) www.3dsystems.com November 2002
(7) www.eos-gmbh.de November 2002
(8) web.mit.edu/tdp/www/ November 2002
(9) www.sandia.gov/media/lens.htm November 2002
(10) www.optomec.com/ November 2002
(11) www.pom.net/ November 2002
(12) www.roeders.com/ November 2002
(13) www.ipt.fraunhofer.de/technologie/prozess/modell/home.html November 2002
(14) www.arcam.com/ November 2002
(15) **Ruder, A., Buchkremer, H. P.,** and **Stover, D.** *Wet Powder Pouring and Rapid Prototyping*, Proc. 1st European Conf. on Rapid Prototyping, Ed. Dickens, P. M., Nottingham, UK, 1992, ISBN 0 9519759 0 0, pp 217–229.
(16) www.wiba.se/ November 2002

R Hague
Rapid Manufacturing Research Group, Loughborough University, UK

3

Rapid Tooling for Magnesium Die-cast Components – An Investigation into Innovative Technologies and Processes for Ericsson Mobile Communications AB

R Hague

3.1 Introduction

In recent years, there has been a proliferation of new techniques to aid in the reduction of product development times and costs. Foremost in these advances have been the developments in the so-called rapid prototyping (RP) techniques. These technologies effectively allow a component to be produced from three-dimensional computer aided design (CAD) data by numerically slicing that CAD file into a series of cross-sections and 'printing' them, one on top of the other, to produce the prototype component – with the method of 'printing' the layers being determined by the RP process that is used. This layer-wise manufacture of components is done in an additive sense, as opposed to the conventional subtractive methods (milling, turning, etc.).

The drawback, with the main RP processes, is that they build parts in materials that are not generally used in conventional products and thus, at present, they can only be considered and used as prototypes. There is much work being currently being performed into new materials for these techniques that will have properties that will eventually enable them to be used as end-use parts – when this happens, prototype components will be able to be produced directly on these additive manufacturing machines and this will lead to a new era of 'rapid manufacturing' (RM). Although there is some limited use of the current machines for RM today, this approach is still some way off.

During the period from now to the time that the RP processes are able to produce parts in the required material, there is a clear need for fast, low-cost tooling solutions to enable companies to produce 'technical prototypes' – which are defined as prototype components that are manufactured in the end-use material. This is especially true of metal components as it should be noted that, the majority of the work currently being performed into rapid prototyping/manufacturing is looking into materials for plastic parts. There is no real viable additive technique for technical metal components – therefore, either investment casting or die-casting techniques are usually employed. For die-cast components, the conventional tooling solutions are historically expensive and have long lead times, making fast tooling solutions attractive.

The advent of the RP processes has led to a marked increase in the possibilities for these fast tooling solutions and thus, the term 'rapid tooling' has become commonplace. However, unlike rapid prototyping and rapid manufacturing, which are defined as being additive manufacturing processes, rapid tooling (RT) can be defined as any method of gaining a tool quickly; indeed, high-speed machining (HSM) is often referred to as an RT technique. However, there is some debate as to the efficacy of some of the available techniques – often some can not be used with any degree of certainty of success, thus leading to a varying degree in uptake and the degree of knowledge available for a given tooling solution when applied to a specific problem.

Most of the work that has previously been conducted into rapid tooling solutions has been for the production of plastic components. The main reason for this is that plastics are simpler materials to mould (especially concerning injection temperatures) and also due to the wider availability of injection moulding machines. Die-casting – in any material – represents a much harsher environment for the tooling inserts and, therefore, proves to be more of a challenge. For example, in injection moulding, the temperature of the material to be moulded would typically be around 250 °C for ABS, whereas, for magnesium die-casting the material is injected at around 650 °C. This increased temperature gives a much greater thermal cycling over a wider range of temperatures and can lead to degradation of the tooling inserts in the form of hot cracking – a phenomena not generally seen in injection moulding.

Additionally, there is a much wider availability of injection moulding facilities around the world than there are die-casting companies. It is also significantly simpler for a research institute to install an injection moulding facility than a die-casting machine. It is these types of places where initial work is often done and, therefore, there are few die-casting facilities installed in organizations that would be best placed to conduct the initial work. Thus, coupled with the fact that the foundry industries (where the majority of the die-casting facilities are held) are historically conservative and are generally less willing to share information has meant that there has been little in-depth work looking into the development of new tooling solutions for die-casting.

Due to the lack of knowledge surrounding the usefulness of the various available RT techniques for magnesium die-casting applications, a unique collaborative project between Ericsson Mobile Communications AB (Sweden), TNO Industrial Technology (Holland), TCG Unitech (Austria), and the Rapid Manufacturing Research Group (now at Loughborough University) was undertaken to investigate potential rapid tooling (RT) methods for the production of prototype magnesium die-cast components for use by Ericsson. The main aim of the project was to significantly reduce the lead-time that it takes Ericsson to produce their prototype die-casting components.

As in all product development programmes, prior to the commencement of this project, Ericsson Mobile Communications AB (hereafter referred to as Ericsson) felt that they would like to reduce the time to produce their prototype magnesium die-castings for use in their mobile communication devices. Typical applications for these magnesium components are for the main chassis (or frame) of a mobile phone. As mentioned previously, die-cast tooling poses a particular problem, as it generally has to be more robust than injection mould tooling in order to withstand the higher temperatures that occur during the die-casting process and therefore, generally takes longer to procure than rapid tooling solutions for injection moulding applications. Therefore, the prototype die-cast components can often produce a bottleneck in the product development process.

Therefore, the main aim of the project was to investigate the various available rapid tooling solutions to see if it would be possible to dramatically reduce the current lead-time for the production of the magnesium die-cast components. These rapid tool inserts should be able to produce magnesium die-cast parts with the same complexity and quality as those produced in conventional die-cast tools. For simplicity, it was considered that the chosen part should have no under-cuts that would necessitate the use of sliding cores in the tool inserts. If necessary, any required undercuts would be post-machined into the die-castings, as is normal for conventional Ericsson parts.

NB: It should be noted that some of the details of the project that are confidential to Ericsson or Unitech have been omitted. In particular, details of cost, lead-time, and specific Ericsson goals are absent.

3.2   Pre-casting work

At the beginning of the project, Ericsson set out a series of requirements for the project. These were to include:

- Tolerances;
- surface quality;
- casting parameters, etc.

3.2.1   Requirements for the project
3.2.1.1   Tolerances
The general tolerances that were decided for the project are detailed below in Table 3.1. It was also stated by Ericsson that smaller tolerances could have been specified on specific functional dimensions for a given feature (depending on the product to be tested), but during the project, this requirement was not necessary.

Table 3.1 Part tolerances

Dimension ranges (mm)			Tolerances (+/−) mm
0	–	10	0.10
11	–	25	0.13
26	–	70	0.15
>70			0.18

N.B. For magnesium, add plus/minus 0.05 for parting surfaces and sliding cores. General tolerance for angles plus/minus 1° for magnesium

3.2.1.2   Surface quality

It was recognized by Ericsson, at one of the project meetings, that the different rapid tooling processes would produce a variation in the surface finish of the die-cast components produced and, indeed, this was one of the factors that was of interest. It was accepted that the quality of the die-cast components should be sufficient to use them as fully assembled technical prototypes and, therefore, it was stated that the inserts should conform to an EDM-type surface finish.

3.2.1.3   Casting parameters

The main aim of the project was to find RT solutions for magnesium die-cast components. Thus, by definition, the material chosen for the tests was the standard magnesium used by Unitech – this being magnesium type AZ 91 HP. The machine identified by Unitech as the most suitable for this type of project was a hot chamber magnesium die-casting machine – a FRECH DAM125. This machine has the appropriate pressure and clamping forces for the relatively small, single impression inserts that would be produced.

3.2.2   Definition of the work part

In order to ascertain which technologies could possibly be suitable for magnesium die-casting, a particular geometry was chosen on which to perform the die-casting trials. It was considered vital that the geometry was representative of the type of part that would be needed by Ericsson for future projects. It was thought prudent that the chosen part should be based on an existing Ericsson design so that direct comparisons could be established. Therefore, the part chosen for the project was the design used for the T28 mobile phone frame. The chosen part is referred to as the 'work part' and a screen shot is shown in Fig. 3.1.

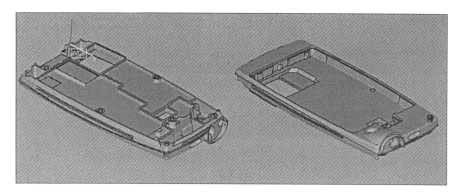

Fig. 3.1 Screen shots of the work part

The advantage of using an existing design was that, as parts were already produced, it was known that the tool *design* works and is able to produce castings and thus, any issues or problems that would occur during the die-casting trials could be attributed to the method in which the tooling inserts were produced.

3.2.3   Tool requirements

In order to facilitate the testing of several sets of inserts from a range of different rapid tooling technologies, it was deemed important to develop a standard bolster system, that could fit on to the selected die-casting machine and into which the various RT inserts could be interchanged and inserted for the die-casting trials. This is a relatively standard approach in plastic injection moulding and is used to reduce costs, as the bolster can often be a significant part of the overall tooling costs. In addition, a crucial element of adopting a standard bolster approach was that the bolster would be made so that it was suitable for future Ericsson projects. The fact that a standard bolster is now available means that there will be considerable time savings on new development projects.

The main areas of interest during this phase were how the proposed rapid tooling inserts would be located within the bolster and the fundamental issue of the gating and ejection systems that were employed. As mentioned previously, the work part was in production at Unitech and thus, existing designs that included information such as shrinkage, gating, and ejection positions, etc., were available. Confidentiality agreements were signed in order to protect the proprietary Unitech information relating to the design of the inserts. However, it was still necessary to decide on the overall dimension of the RT inserts.

Again, for ease, the bolster design was based on one that was already in existence at Unitech. However, in order to make the tooling arrangement useful for future Ericsson projects, the tooling solution that was developed concentrated on splitting the design into three different areas. These included the following.

- *Bolster and primary gating* – The bolster and primary gating were produced to fit the FRECH DAM125 hot chamber die-casting machine as identified as the most suitable for this particular project. The bolster produced is independent of the part to be die-cast.
- *Frame* – The frame is in effect a 'mini-bolster' whose main function is to minimize the insert dimensions – this was because for many of the rapid tooling technologies, there is a maximum build volume that is available. The frame was manufactured for a specific insert dimension and also included the heating channels. The frame fits into the bolster – described above.
- *Inserts* – The inserts contain the part geometry and also the secondary gating. They also contained the geometry from where the ejection of the part is effected. The inserts were fitted into the frame. The inserts were single impression only.

A schematic diagram of the tooling arrangement is shown in Fig. 3.2.

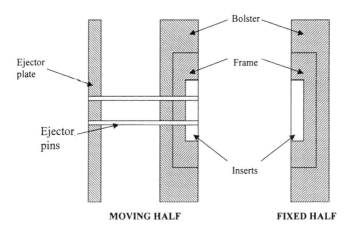

Fig. 3.2 Schematic of the proposed tooling arrangement

3.2.4   Review and selection of potential tooling solutions
As described in the introduction, the term 'rapid tooling' can be used to describe any method of obtaining a tooling solution quickly. In summary, this can be by three main methods.

- Subtractive rapid tooling (i.e. machining).
- Direct 'RP' type rapid tooling (additive/layer-wise).
- Indirect rapid tooling (by the use of a master pattern).

After careful consideration of the most appropriate RT solutions that were on the market (or near to release) at the time of the project, the following technologies, as listed in Table 3.2, were selected for inclusion in the current project along with the vendor considered best equipped to supply the inserts. It should be noted that this Chapter is not intended to constitute a review of the different RT solutions listed. However, a fuller description of the RT techniques listed in Table 3.2 is given in Chapter 2

Table 3.2 Selected RT technologies

	Process	Vendor
1	HSM of STL files	TNO
2	Direct metal laser sintering (DMLS) (50 µm system)	RMRG
3	DMLS (20 µm system)	RPI, Finland
4	Laserform ST100	Land Rover limited
5	Laser caving	TNO
6	ProMetal	Extrude Hone
7	Keltool process	3D Systems
8	WIBAtool process	WIBA
9	Direct metal deposition (DMD)	POM
10	Laser engineering net shape (LENS)	Optomec
11	CMB	Röders
12	Arcam	Arcam

3.2.5   Design of the work part inserts and mould flow

Having decided on the work part, it was then necessary to produce the CAD data for the required core and cavity inserts that would be produced by the various RT processes listed in Table 3.2.

The insert pair were designed by Unitech using the SolidEdge CAD system based on the original Unigraphics file of the work part supplied by Ericsson. The insert design was based on their previous experience of the work part and also in accordance with the specification previously laid out. No mould-flow analysis was performed because this was an existing tool design that was successfully in production and, therefore, mould-flow was deemed unnecessary.

Figure 3.3 shows the design of the inserts in the as-finished state. The files shown are screen shots of STL files. It should be noted that it was not possible for some of the rapid tooling vendors to use the STL format – in these instances, IGES files of the inserts were sent.

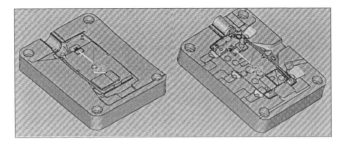

Fig. 3.3 Insert pairs of the insert pair

3.2.6 Production of the bolster

Unitech produced the bolster to the specification laid out. In addition, as TNO were to be responsible for the preparation of the inserts for the die-casting trials, a 'gauge' – with exactly the same dimensions as the production bolster and also including ejector pins and other location holes – was produced to aid them in the fitting of the inserts. This was to ensure the minimal amount of work when the inserts were sent to Unitech for fitting into the production bolster. This gauge-set is shown in Fig. 3.4.

Fig. 3.4 Gauge used by TNO for accurate fitting of the inserts
(this use of the gauge proved to be an extremely useful aspect of the project and greatly aided in the fitting work on the inserts)

3.2.7 Production of the work part inserts

For all the 12 RT technologies, a vendor was chosen. However, out of the 12 inserts that were chosen for the trial it was only possible to have nine produced. Despite the best efforts, it was not possible to get inserts made in:

- LaserCaving;
- controlled metal build-up (CMB);
- ARCAM process.

LaserCaving – Despite being a very interesting technology that should be capable of producing fine detailed accurate tooling inserts, the LaserCaving technology is not, at the time of writing, in a developed state to be included in this project.

CMB – It was not possible for Röders to produce the inserts in the CMB process because, at present, the system is not technically developed enough.

ARCAM – Due to the method of material processing, the Arcam process is unlikely to be able to produce the detail necessary either in the medium- or long-term without further finishing of the tools. However, it should be noted that Arcam position the process as a production tooling rather than rapid tooling solution.

3.2.8 Fitting of the work part inserts into the bolster

Upon receipt, the inserts were prepared to make them ready for the die-casting trials to be held at Unitech. (This fitting work was done with the use of the gauge, as detailed in Fig. 3.4.) The fitting work included:

- drilling and reaming of ejector pin holes;
- ensuring adequate shut-off of the parting surfaces;
- milling the outside edges and back to fit the bolster.

3.2.9 Verification of the work part inserts

For the measuring (and machining) of the inserts, Unitech provided TNO with two-dimensional drawings that included critical dimensions, and ejector mounting locations. During one of the project meetings, critical dimensions were located that would be used to measure the inserts relative to the drawing and each other. Table 3.3 details the dimensions that were ascertained.

Table 3.3 Critical dimensions of received inserts

Position	Bottom			Top			Conclusions
	P1	P2	P3	P1	P2	P3	
Drawing	4.62	68.42	13.46	10.20	78.00	1.26	
HSM	−0.02	+0.03	−0.05	−0.08	−0.07	−0.01	'Good'
DMLS 20	−0.03	−0.21	+0.01	+0.04	−1.22	+0.17	Out of tolerance
DMLS 50	−0.04	+0.26	+0.06	−0.08	+0.06	0	Out of tolerance
Laserform ST100	−0.04	+0.16	+0.49	−0.02	+0.18	−0.01	Out of tolerance
ProMetal	+0.01	−0.05	+0.40	−0.66	−1.02	−0.02	Out of tolerance
Keltool	0	−0.02	+0.13	+0.01	+0.04	−0.01	Out of tolerance
WIBAtool	−0.64	−0.01	+0.76	−0.07	−0.07	+0.04	Out of tolerance
LENS POM	–	–	–	–	–	–	Out of tolerance
LENS Optomec	–	–	–	–	–	–	Out of tolerance

It should be noted at this point that the inserts from the two LENS processes (Optomec and POM) were impossible to measure as they were, in-effect, 'near-net shape processes'. The decision was taken not to attempt any form of finishing on the inserts and thus, casting trials were not performed.

3.2.10 Casting parameters

The following are the general parameters that were used, for all the casting trials, on the FRECH DAM125 Hot Chamber magnesium die-casting machine.

Magnesium melt temperature = 640 °C
Tool temperature   = 200 °C

3.2.11 Casting procedure

The procedure, for all of the tests that occurred, was that the inserts were placed in the heated bolster (on the die-casting machine) and were left, for about 30 minutes, to reach near to the operating temperature. The inserts were then given a heavy lubrication coating that was treated with a blowtorch. The first shot was applied and, if successful, was discarded as it contained the detritus of the heavy lubrication. This was repeated until the castings were free of dirt. The casting trials then continued using the parameters listed above and the castings were numbered and separated.

3.3   Results and discussion of casting trials

3.3.1   HSM of STL files

It was considered important that some reference point to the entire trials should be done by performing the first test on the inserts thought most likely to achieve viable casting. Therefore, for these first trials, the inserts that were produced with HSM from STL files were tested, as they were considered by the group to be the most likely to work.

The casting trials for the HSM inserts was executed, as described above. As expected, the inserts were able to produce the required castings. There was no sign of degradation in the tool and no sign of hot cracking. Fig. 3.5 shows some of the castings achieved from the HSM inserts.

Fig. 3.5 Castings from HSM inserts
(it should be noted that, the 'flash' surrounding the castings is normal in magnesium die-casting and is not particularly a function of a bad parting surface)

3.3.2   DMLS 20 μm

The DMLS 20 μm inserts were then prepared and cast. Again, five hundred castings were produced without any serious degradation in the tool, apart from a small pin feature that was removed after a couple of hundred shots. Figure 3.6 shows a casting from the DMLS 20 μm insert before it has been ejected.

Fig. 3.6 Shots of DMLS casting before ejection

3.3.3   DMLS 50 μm

After heating-up the bolster/insert, the insert was sprayed with lubricant and heated separately using a hand-held gas burner. The first casting was attempted and upon opening the tool, a section of the cast part was stuck in the cavity side with the remainder also being stuck on the core-side. It could be seen that, during the ejection phase of the casting cycle, the ejector pins had actually penetrated (punctured) the cast part so that it was not ejected. This is demonstrated in Fig. 3.7.

Fig. 3.7 Casting stuck on DMLS 50 μm inserts and rough surface finish

On initial inspection, it appears that the DMLS 50 μm inserts (which were unfinished, apart from a light shot-peening) were far too rough. Because of this, the casting 'gripped' around the roughness and froze on to the tool. This roughness is demonstrated in Fig. 3.7 – it can also be seen that part of the casting is stuck in this side of the insert. The casting trials on the DMLS 50 μm inserts was terminated at this point.

3.3.4 ProMetal

The ProMetal inserts were mounted to the bolster on the machine and after lubrication and heating the first shot was made. Again, the casting stuck on the core-side of the tool with the ejector pins puncturing the casting similar to that of the DMLS 50 μm inserts.

3.3.5 Keltool

After the inserts were mounted heated and lubricated, as in previous casting trials, the first shot was made, but, during ejection some pieces of the part stuck in the core-side – this was because some areas of the tool had not been finished enough. It was decided to leave the inserts on the machine and gently grind the very minor areas that needed finishing with a diamond hand-grinding tool. Following this, 500 castings were successfully die-cast. The following are points to be raised from the trials.

- There was no evidence (by checking the castings or inserts by sight) of hot-cracking.
- At part number 355, the small pin-detail found in the cavity side broke off and stuck in the casting. This pin feature is shown in Fig. 3.8.

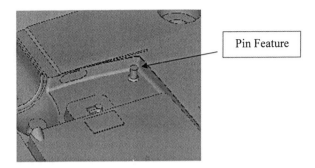

Fig. 3.8 Pin feature removed during casting

Figure 3.9 shows one of the castings achieved from the Keltool inserts after hand-polishing had been applied.

Fig. 3.9 Casting from Keltool inserts

It was clear from this set of casting trials that a good surface finish is crucial to the success of the inserts for casting.

3.3.6 WIBAtool

For these inserts, it was possible to gain the required number of castings. However, for the first time in the trials, the first signs of hot-cracking appeared at around 90 shots and got progressively worse until the final casting was made. The hot-cracking in the inserts produced defects in the subsequent castings and this is shown in Fig. 3.10 for casting number 135.

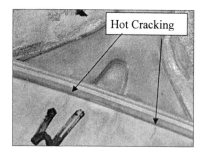

Fig. 3.10 Emergence of hot cracking in the WIBAtool castings

The following are the main points to arise from the casting trials on the WIBAtool inserts.

- Castings achieved.
- First signs of hot-cracking occurred around the gating area, at about 90 shots.
- At 150+ shots, the hot-cracking was more visible and also, there was increasing numbers of cracks (visible in the tool and the resulting castings).
- Hot-cracking increased in severity, but the tool did not catastrophically fail.
- The small pin feature on the cavity insert, however, broke off and stuck in the casting at shot number 452.

3.3.7   Laserform ST100

The Laserform ST100 inserts were loaded on to the machine. The core and cavity were lubricated and heated and the first casting was attempted. The result was the same as for the DMLS 50 μm inserts with the part being stuck on the inserts with the ejector pins puncturing the component. The part was removed from the tool and another shot was tried with the same effect. Again the process was repeated, with the same result. Figure 3.11 shows the casting stuck on the Laserform ST100 insert.

Fig. 3.11 Casting stuck on Laserform ST100 insert

Additionally, with the Laserform ST100 inserts, it appears that the infiltrant that is used during their manufacture had partially melted – indicating that the inserts may not be suited to die-casting due to the higher temperatures involved. This is shown in Fig. 3.12.

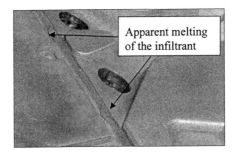

Apparent melting of the infiltrant

Fig. 3.12 Apparent melting of the Laserform ST100 infiltrant

It could also be observed that while trying to remove the stuck castings from the tool that it was extremely well stuck to the insert. Discussions with metallurgists suggest that the molten magnesium is dissolving the tin within the bronze infiltrant of the Laserform ST100 and bonding with it. It is this that is causing the strong adherence of the cast part to the Laserform ST100 insert. This is shown in Fig. 3.13.

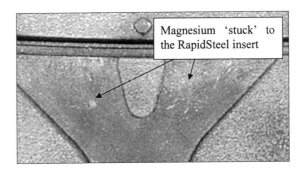

Magnesium 'stuck' to the RapidSteel insert

Fig. 3.13 Adherence of magnesium to the Laserform ST100 material

3.4 Measurement and testing

3.4.1 Measurement of inserts after die-casting trials

Table 3.4 shows the measurements of the critical dimensions on the inserts after they had been through the die-casting trials. Although the Keltool and WIBAtool produced castings, these castings were out of tolerance due to the tools that they were produced in – this is most likely due to the 'indirect' method of production for these two tooling types.

Table 3.4 Measurement of critical dimensions after casting

Position	Bottom P1	P2	P3	Top P1	P2	P3	Conclusions
Drawing	4.62	68.42	13.46	10.20	78.00	1.26	Good
HSM	−0.03	+0.04	+0.07	−0.16	−0.06	−0.01	
DMLS 20 µm	−0.02	−0.21	−0.02	−0.03	−0.08	0	Good

3.4.2   Manufacture of functioning phones

3.4.2.1   *Post processing work*

Although not required in the original project specification, after the die-casting trials were performed, a certain number of the castings that were produced on the HSM and DMLS 20 µm inserts were prepared in the standard method to enable them to be built into a functional mobile phone that would allow them to be subjected to Ericsson drop tests. The operations that were applied included:

* machining of undercuts;
* painting;
* application of gasket.

Figures 3.14 and 3.15 show an example of the DMLS and HSM casting (respectively) that went through the post-processing stages.

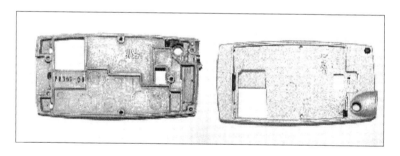

Fig. 3.14 DMLS Casting (No. 400) after post processing

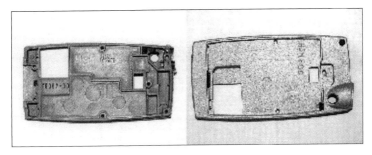

Fig. 3.15 HSM Casting (No. 500) after post processing

3.4.2.2   Final manufacture

After post-processing, the built-up frames were sent to Ericsson to be assembled into functioning mobile phones so that they could be subjected to the standard Ericsson drop tests. Figures 3.16 and 3.17ow DMLS and HSM frames (respectively) that have been assembled into functioning T28 mobile phones.

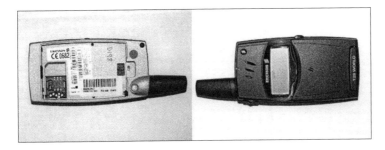

Fig. 3.16 DMLS frame built into functioning T28 mobile phone

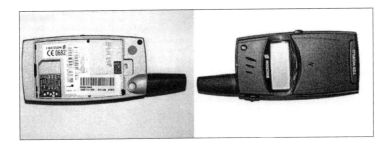

Fig. 3.17 HSM frame built into functioning T28 mobile phone

3.5   Discussion

3.5.1   Summary of results

An overview of the rapid tooling inserts that were investigated from the point where they were identified as being potential tooling solutions to the production of successful castings is shown in Table 3.5. In summary, only the HSM of STL files and the DMLS 20 µm systems were able to achieve successful castings.

Table 3.5 Summary of the inserts investigated to successful castings produced

Potential solutions	RT Inserts produced	Castings attempted	Castings produced	500 successful castings
HSM DMLS 20 µm	HSM DMLS 20 µm	HSM DMLS 20 µm	HSM DMLS 20 µm	HSM DMLS 20 µm
Keltool WIBAtool DMLS 50 µm ProMetal Laserform ST100 DMD POM LENS Optomec Laser Caving CMB Arcam	Keltool WIBAtool DMLS 50 µm ProMetal Laserform ST100 DMD POM LENS Optomec – – –	Keltool WIBAtool DMLS 50 µm ProMetal Laserform ST100 – – – – –	Keltool WIBAtool – – – – – – – –	– – – – – – – – – –

The main aim of the project was to design the tooling inserts, produce them on the appropriate technology, fit them into the bolster, and then perform the die-casting trials in a reduced timescale. Some of the RT technologies have the potential of significantly reducing the lead-times over the time that it currently takes Ericsson to produce their prototype magnesium die-castings – therefore, representing a major step forward.

3.5.2   Rapid management

To truly achieve the main aim of reducing time, it is not just a matter of designing the inserts, manufacturing them, and then performing the die-casting. Clearly, not all this time is taken up in insert design, machining, or EDMing of the inserts, etc. – tools have to be scheduled to fit in with other work commitments. If a reduced time is needed then some kind off 'rapid management', 'rapid scheduling', and 'rapid decision making' is essential – this is the same for all companies that require a rapid turnaround. The manufacture of the inserts is obviously important, however, if there is no capacity on the required HSM or DMLS 20 µm machine, then what will happen to the schedule? Therefore, it is considered that the management and scheduling of the prototyping route and not just the choice of RT technology, is one of the key methods of attaining a reduced lead-time.

3.5.3   Future vision

The results of this case study have shown that 'rapid tooling', as a group of technologies, is struggling to keep up with HSM. Some of the technologies tested within this project will have a limited lifespan – in general, it is probably worth avoiding indirect tooling methods that require a master pattern as they are generally unable to hold the dimensional tolerances necessary. This has been demonstrated within this project with the Keltool and WIBAtool processes.

The application of the most appropriate tooling solution to the task in hand is the key to success – several of the tooling solutions that were tested and failed during this project are more than suitable for plastic injection moulding, however, they were unable to cope with the rigours of die-casting.

3.6   Summary

Out of the 12 'appropriate' RT technologies that were globally sourced, only two were able to produce viable castings. The successful RT processes were HSM of STL files and the EOS direct metal laser sintering (DMLS) 20 μm system. Both processes were able to produce within-tolerance castings that could be used for the production of aesthetically acceptable, functional mobile phones that were capable of withstanding the standard Ericsson test procedures.

Some of the technologies investigated were unable to either be manufactured or to produce castings. However, there is still some room for development in some of these, and further tests should be performed when the technologies become viable. It should be noted, however, that – in a world of rapid change – that some of the tooling technologies that failed to perform as expected within this study might have improved since the time of the project and thus, the results given should only be taken as indicative. However, it should also be noted that rapid tooling in general is struggling to keep up with the advances in HSM and it would take significant advance to overtake this long established and continually developing technology.

Acknowledgements

The author would like to thank the project partners from Ericsson Mobile Communications (now part of Sony Ericsson), TNO Industrial Technology, and TCG Unitech for their great contributions and joint leadership of this study.

R Hague
Rapid Manufacturing Research Group, Loughborough University, UK

4

Rapid Tooling for Aluminium Die-cast Components – An Investigation into EOS DMLS 20 μm Tooling Inserts for a 'Clutch Housing' for Dyson

R Soar

4.1 Introduction

EOS (Electro Optical Systems) GmbH are constantly undertaking field trials of their new materials. In line with this effort, the Rapid Manufacturing Group became involved in collaborative research to test EOS's 20 μm DirectSteel material, for trials in the area of high-pressure die-casting of aluminium LM24. This application is demanding, for any rapid tooling process, and would represent a move forward by EOS, whose previous materials were intended for injection moulding applications.

A twin cavity 20 μm DirectSteel tool was built based on the design of an existing vendor's product. This product was a clutch housing mechanism for a Dyson vacuum cleaner. The original tool was built as a three-part insert bolted into a production bolster. The inserts had already been run over many tens of thousands of castings by Kemlows Die-casting Limited and it was for this tool that one set of the inserts (core and cavity) were produced in 20 μm DirectSteel by EOS. These were set in the tool and mounted back on the same die-casting machine and run by Kemlows Die-casting Limited.

4.1.1 Aluminium die-casting

Aluminium die-casting is an important technique for the mass production of near net shape components and is still the major automotive casting route to lightweight components for use in stressed areas. Of the many casting techniques available, the high-pressure die-casting of aluminium components is the most arduous application due to the pressures and temperatures required to produce parts. The high-pressure die-casting process produces the lowest cost per part for subsequent castings but requires the highest level of capital investment due to the complexity and longevity required of the tooling.

As aluminium components are increasingly required for more stressed applications (engine/powertrain components) so the demands on the process have also increased. Die-casters must compete with the increasing application of reinforced polymer components being developed for similar applications. Tied to the move of die-casting and toolmaking to the Far East, then there is a sense of urgency among the high-pressure die-casting toolmakers who urgently require rapid tooling solutions that will give a faster return on a tool. Such tools may not be required to produce 100 000s of castings but possibly as little as 1000s, where short run tooling is required, or even 100s of castings, where technical prototype/bridge tooling is required, for die design validation, short runs, etc.

4.1.2   DMLS 20 μm DirectSteel tooling

Various rapid tooling solutions are currently available, or under development, for a wide variety of applications. For pressure die-casting applications only a few rapid tooling solutions exist due to the temperatures and pressures that a tool is expected to endure. In fact, at the time of this research, the only rapid tooling solutions that existed for aluminium high-pressure die-casting included laminate tooling and EOS's direct metal laser-sintering (DMLS) process using DirectSteel 20 powder built in 20 μm layer thicknesses.

The 20 μm DirectSteel process relates not to the powder size (though the powder is finer than the 50 μm system) but to the layer thickness that the DirectSteel powder is deposited in the EOSINT M 250 *Xtended* system. The DMLS process is technically referred to as a liquid phase sintering process, whereby the powder is selectively heated by an infrared CO_2 laser beam. As the powder heats up a low-melt constituent melts and binds to those particles in the powder that have a higher melting point. During this process densification occurs as the powder solidifies.

By using finer powder and thinner build layers it was hoped that this powder would produce a better surface finish and offer more robust parts than the 50 μm DirectSteel process. This study attempts to test the material for use in the production of aluminium high-pressure die-cast components.

4.2   Aims and objectives

After initial discussion with the Project partners the aims were as follows.

"To investigate the feasibility of using the DMLS 20 μm DirectSteel tooling system for the high-pressure die-casting of 500 aluminium components that would satisfy technical prototypes, bridge, and short run production tooling." In addition to this aim, the project also sought to "Consider the implications when designing DMLS 20 μm DirectSteel tools for this application". Therefore, following on from the project goals, in summary the objectives of the project were:

• to move quickly from CAD model to production tool;

- to identify whether 20 μm DirectSteel tooling can be used for high-pressure die-casting of existing aluminium components;
- to identify whether the tool could produce 500 castings to satisfy technical prototypes, bridge, and short-run production tooling;
- to highlight any issues specific to 20 μm DirectSteel tooling in this application (build time/sealing/finishing/machining);
- to study the quality/tolerances of any subsequent castings;
- to identify the costs of this process.

4.3   Methodology

From the outset it was clear that whatever die design was chosen to build a 20 μm DirectSteel tool from, then the tool should be run under full production conditions and ideally be a duplicate of an existing component as a form of comparison. Subsequently, conversations with one of the partners, Kemlows Die-casting Limited, regarding a suitable component, revealed that they had been producing a clutch housing component for the Dyson vacuum cleaner range. This component had a multi-insert bolster assembly and was required to work at the maximum operating conditions to ensure that Kemlows could recoup the costs on such a low-cost part (£0.13 per casting). Dysons were receptive to the idea and agreed to allow one of the two cavity inserts to be duplicated as a 20 μm DirectSteel tool.

A copy of the engineering drawing of the core die is shown in Fig. 4.1. The drawing shows the core (or moving) side of the die including the three inserts (numbered 1–3) set into the bolster. Two of the inserts form two of the twin cavity inserts and the third is a cooled runner insert.

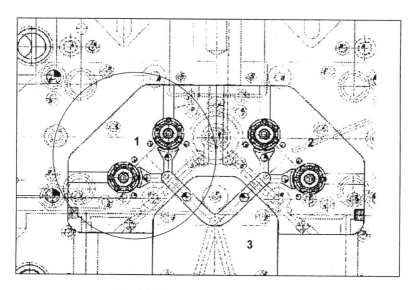

Fig. 4.1 Engineering drawing of core die

4.3.1   Design of the 20 μm DirectSteel inserts

Dyson supplied the Unigraphics three-dimensional CAD file of the clutch housing in order that inserts could be made (as shown in Fig. 4.2).

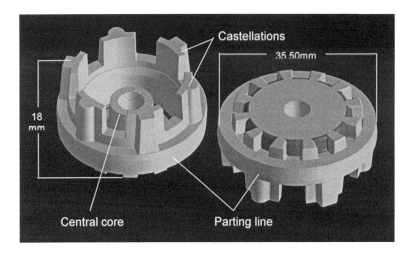

Fig. 4.2 UG CAD model of clutch component

From the CAD model of the part, a 0.6 per cent shrinkage factor (as specified by Kemlows for LM24 alloy) was applied and the split line defined. This model was then used as a Boolean operation to form the male and female half of the tool. The subsequent STL files for the inserts are shown in Figs 4.3 and 4.4.

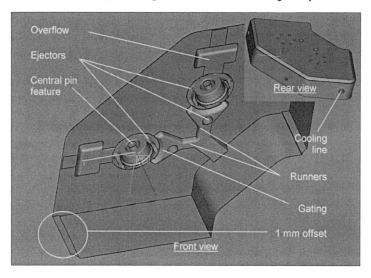

Fig. 4.3 CAD model of core (moving/ejector) side of 20 µm DirectSteel insert

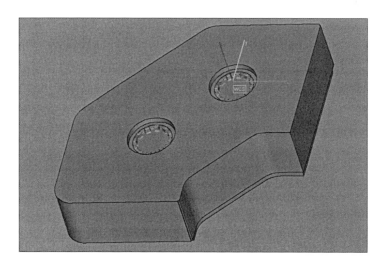

Fig. 4.4 CAD model of cavity (fixed) side of 20 µm DirectSteel insert

Figure 4.3 shows the core (or moving/ejector) side of the die insert including runner, gating, and ejectors. The diagram also shows the inclusion of cooling/heating lines as well as a recess to pass a thermocouple into the centre of the tool. Ejectors were located to act against the runner and each of the castellations on the component. In addition, a 1 mm offset was applied to the external faces of the inserts for machining back. From previous experience, overflows were required to remove airlocks and impurities that can form on the flutes and castellations that form the clutch housing.

Figure 4.4 shows the cavity (fixed) side of the insert. Again a 1 mm offset was applied to ensure an accurate fit into the bolster after machining back. Once built as a DMLS 20 μm DirectSteel insert, a BSP threadform would be machined to form the connection to either heating lines to bring to the tool to casting temperature (200 °C) or water cooling lines during the run.

Of note are the central pins in the design (see Fig. 4.3). A two-stage approach was decided upon to test the inserts in the die-casting environment. In the production Dyson tool these central pins are formed by nitrided pins set into the tool to form a shut-off for a spindle that passes through the clutch housing to the motor unit on the vacuum cleaner. To attempt to reduce the time to produce and run the inserts it was decided to build the central pin features during the DMLS process. Although this approach raised doubts about the DMLS pin's ability to resist the process of metal injection, shrinkage, and then part ejection, it was felt that this should be explored to identify possible design limitations for the DMLS process. If this pin should fail during the first run then a solid nitrided pin would replace this feature, as in the existing tool.

4.3.2   Building the 20 μm DirectSteel inserts

The Unigraphics CAD models of both halves of the insert were converted to the STL format. The STL format is used by all current rapid prototyping (RP) technologies and defines a solid model as a series of connected triangles (tessellations) that are formed over its surface. As each 20 μm slice or layer is projected through the tessellated model the program can then identify what part of each slice is solid material (to be scanned with the laser) and which is surrounding void (to be left as loose powder).

As stated in the product name, the EOS 20 μm DirectSteel process builds each layer of the object to be built by spreading 20 μm of material on to a solid base plate. As each layer is deposited and compacted it is scanned by a 200 watt CO_2 laser to melt the powder as defined by the slice file. The next layer is then spread and the process is repeated until the object is completed.

Building in 20 μm layers ensures a far higher degree of finish (and less obvious stair stepping) compared to the previous DMLS 50 μm process but does also imply that the process takes longer to build a similar component. For this reason EOS stress that it is important to minimize the size of the part as much as possible to ensure valuable build-time is not wasted on a tool insert. In the case of the Dyson insert only one half could be built inside the build volume of the machine and consequently took 179 h (83 h for core and 96 h for cavity) to build the insert pair.

Had the component been optimized (i.e. built as a smaller component) then EOS estimate the build-time would have come down to 80–100 h for both halves. Reducing the insert's size was considered by the group but would have involved the production of a secondary clamping jig (mini-bolster) that would have added time to the process as it was the group's intention that the inserts should fit straight into the Dyson bolsters. EOS offer advice on clamping systems so that smaller parts can be built to reduce costs.

One area where savings on build-time could be made was in the depth of the tool that was built. The EOS process uses a solid-metal base plate that serves two purposes. The first serves to absorb some of the energy of the laser scanning over its surface to bond the first layer of powder to its surface. The second serves to minimize any distortion that can occur in the part due to the rapid heating and cooling of the part as the laser fuses the powder to form the solid object.

Having a solid base plate (build platform) also provides a strong base that the wiper blade can compact the powder on to as it sweeps each consecutive layer on to the layer below. By increasing the thickness of the base plate used then a direct saving in build-time is achieved in the 'z' axis. In the case of the inserts a substantial base plate could be used to save building layers up to the level of the cooling lines defined through the tool. Figure 4.5 shows the STL file prior to slicing as well as the reduced thickness of the core side of the tool.

A further consideration in the construction of all EOS metal parts is the build strategy used to produce the tool. As stated previously, the key to effective use of the process is saving time when building the part. Where large solid volumes are required then the process has the ability to speed up the build through a strategy called 'skin' and 'core'. 'Skin' relates to the build strategy used to build all those features that form the external surfaces or skin of the part. It is the surfaces that must endure repeated casting forces and are, therefore, built with a scanning pattern that produces a near fully dense surface. Skin depth can be varied but normally is in a range of 3–5 mm deep.

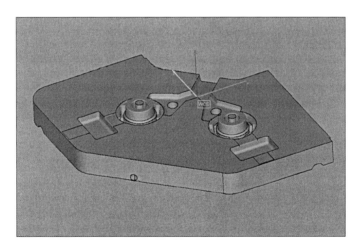

Fig. 4.5 Modified STL model for minimized build-time

The core strategy relates to the internal volume of a part and uses a reduced 'dwell' time with the laser that enables the laser to scan much faster over these sections of a layer. The core strategy results in a porous structure of about 70–80 per cent density.

One consideration for the Dyson inserts was the inclusion of ejectors and cooling lines. In the construction of a conventional tool, these are drilled after the tool cavity is formed. In the case of a 20 μm DirectSteel build then all ejectors and cooling lines must be included in the initial CAD model so that these features are built with the skin strategy. If the cooling lines are drilled after the parts are built, for example, then coolant would be forced into the porous body of the tool (this is being investigated for its positive connotations). Where fluid passes through the inserts the walls must be fully dense to prevent leakage.

The STL files were sent to EOS in Finland where they built the two halves of the insert. At this stage, the 20 μm DirectSteel process is not capable of producing surfaces smooth enough to ensure immediate ejection of die-cast parts. Typical surface roughness for unfinished components is 3.0–4.0 microns R_a.

To ensure ejection of the castings, and to prevent a casting freezing to the central pin feature (as previously described), secondary hand finishing was applied to the inserts. In addition proprietary shot peening was applied to the inserts before and after hand polishing to further smooth and compact the surface to withstand the die-casting process. This was performed by RPI Finland to 0.5–1.0 microns R_a. The completed inserts are shown in Fig. 4.6.

Fig. 4.6 Core and cavity DMLS DirectSteel inserts

4.3.3   CMM comparison of inserts versus CAD model

The inserts were then passed to Dyson who produced a co-ordinate measuring machine (CMM) report as a comparison against the dimensions of the CAD model of the inserts. Though not critical for this experiment (as the inserts would have secondary finishing prior to insertion in the Dyson bolster), the inserts were measured for accuracy/tolerance so that this data could be combined with data generated from similar projects as a means of reference.

Dimension for the two halves of the tool are shown in Figs 4.7 and 4.8.

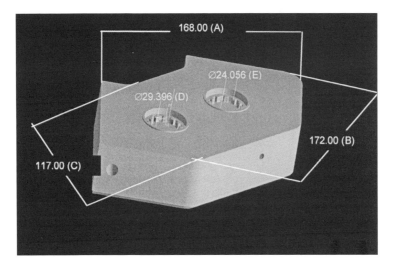

Fig. 4.7 CMM dimensions used for cavity insert

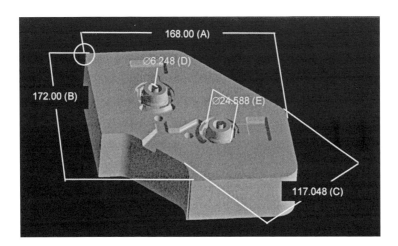

Fig. 4.8 CMM dimensions used for core insert

Measurement was performed on Brown & Sharpe (Mistral 10.07.07) CMM machine with a measuring accuracy guaranteed to ±0.010 mm. Dyson were looking for tolerances of ±0.15 mm on all dimensions and Table 4.1 shows the results of this exercise.

Table 4.1 CMM accuracy data for core and cavity

Measurement	Difference	Type	Nominal	Actual
Cavity				
A	− 0.231	Dimension	168.000	167.769
B	+0.037	Dimension	172.000	172.037
C	− 0.086	Dimension	117.048	116.962
D_1	− 0.11	Dimension	29.396	29.286
D_2	− 0.091	Dimension	29.396	29.305
E_1	− 0.15	Dimension	24.056	23.906
E_2	+0.035	Dimension	24.056	24.091
F	+0.11	Flatness	0.000	0.110
Core				
A	+ 0.13	Dimension	168.000	168.130
B	−	Dimension	172.000	172.000
C	− 0.108	Dimension	117.048	116.940
D_1	+0.04	Dimension	6.248	6.252
D_2	− 0.02	Dimension	6.248	6.246
E_1	− 0.002	Dimension	24.588	24.586
E_2	+0.02	Dimension	24.588	24.608
F	+0.12	Flatness	0.000	0.120

Dysons were able to confirm that all dimensions except dimension 'A' on the cavity inserts were within tolerance. However, this particular part was to be post machined and, for this dimension, accuracy was not critical.

4.3.4 Assembling the tool

As stated previously, the 20 μm DirectSteel inserts made up only one set of the two twin cavity inserts in the bolster. In the initial stages of the project options such as producing one set of inserts with a shot peened finish only were considered. This was subsequently discounted as, if one set of inserts should fail it may shorten the life of the other. Using just one set of 20 μm DirectSteel inserts would ensure that the job of setting the new inserts in the tool would be easier as much of the existing tool was still in place.

The 1 mm offsets were milled off the sides and the parting plane of the inserts dressed. The ejectors and BSP threadforms were then applied for the cooling lines. As a matter of course, the toolmaker performed a pressure test on the cooling lines before releasing them to Kemlows. During this process extensive leaks appeared around the inserts. Of note is the fact that even though Kemlows had difficulty producing the threadforms the BSP fittings did not leak.

A meeting was held with EOS to try to pin down where the leaks were most likely to be found and they were able to identify two possible causes. The first was the interface between the base plate and the DirectSteel material. Due to the stresses (or sometimes impurities left) on the base plate it is possible that the first layer did not adhere sufficiently. In the case of the 20 μm DirectSteel inserts the base plate had been used to replace the building of much of the thickness of the tool and this had brought the base plate very close to the cooling line. Water may have been leaking at this interface.

The second possible cause was leakage through the walls of the cooling lines themselves. Though the skin strategy is designed to produce near fully dense external surfaces, EOS had noted problems when producing down facing surfaces, particularly where they make up a skin over unfused powder as was the case with the cooling line (possibly down to 80 per cent of full density). Where solid sections are produced, then the compressed and solidified material beneath that layer will normally ensure a dense skin. However, where a skin is formed over loose powder then compaction by the sweeper blade can be impeded.

Under injection moulding conditions EOS would normally solve this by partially infiltrating the cooling lines with an epoxy resin. In die-casting average die temperatures (200 °C) are very close to the point at which epoxy resins break down. Even so, it was decided to pursue high-temperature epoxy infiltration based on the fact that if the epoxy were capable of sealing the cooling lines then even if the tool heated up beyond the point the resin breaks down then it may still be able to seal the cooling lines. Indeed, it was speculated that even if the epoxy resin was to burn inside the tool then a deposition of carbon would be left in the voids that could possibly still seal the tool. This concept would later be tested as the same cooling lines would be used with the oil heating unit (working up to 500 °C) to bring the tool rapidly up to casting temperature.

The assembled tool is shown in place in Figs 4.9 and 4.10 on the Frech 125 die-casting machine that was used for the trials.

Fig. 4.9 Frech 125 tonne high pressure die-casting machine

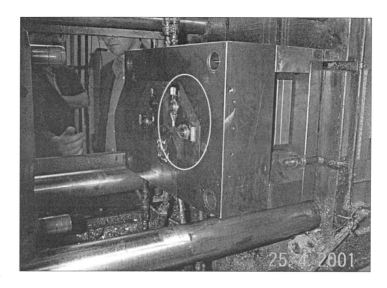

Fig. 4.10 Core 20 μm DirectSteel insert (circled)

4.4   Experimental procedure

As outlined in the objectives, the tool was to be run at full production pressure, injection velocity, and handling strategy (ejection, part removal, and ladling) for as many castings as possible up to the goal of 500 castings or whenever the tool failed or was unable to produce castings of sufficient quality as deemed by Dyson (who observed the run).

It was decided that to minimize the shock to the inserts a high-temperature grease would be applied to the inserts for the first few castings. This is standard procedure dependant on the type of release agent system used. Such greases normally contain molybdenum or graphite and ensure part removal even though they will give a 'dirty' casting. After ten shots the temperature of the tool should normalize and casting can continue with the standard polysiloxane lubricants (in this case ChemTrend's Safetylube 3180) that will give a clean casting with minimum loss on tolerance.

During the run, five castings in every 50 (i.e. 0–5, 45–50, 95–100 etc.) were kept for inspection for:

- drift from tolerance in the castings over the run;
- breakdown of tool by observation of defects or witness marks.

The first run would be performed as long as the 20 μm DirectSteel central pin feature stood up to the casting process. Machine parameters were not changed from those used with the original Dyson tool as it has been dedicated to the production of the Dyson components for some time. Machine parameters were:

- cycle time 18 seconds;
- injection velocity 40 m/s;
- cavity fill time 30 milliseconds;
- plunger velocity of 1.2 m/s.

4.5   Results

4.5.1   Results from first run

A total of 41 shots were taken from the twin cavity 20 μm DirectSteel insert before the run was terminated.

For all of the castings in the first run, aluminium froze to the central pin feature on the first casting but did not shear the central pin off. Instead the casting sheared and this allowed the run to continue with the central pin section still in place. The sheared casting is shown in Fig. 4.11.

Fig. 4.11 Sheared central pin on casting

At castings 8 and 14 the run was paused to attempt to release the sheared section by chiselling it out of the frozen casting around the central pin feature. This was not possible and it was decided to try to 'cast out' the problem.

At casting 35 the run was paused to inspect the tool for hot cracking/soldering on the surface. None was visible.

It should be noted that horizontal lines were observed on the castings as shown in Fig. 4.12, and this was explained by EOS as being caused by a fault on the scanner when the DMLS inserts were being made.

Fig. 4.12 Strata lines (circled) due to stoppage in DMLS build

The run was terminated at 41 castings primarily because it was felt that to demonstrate the DMLS 20 μm DirectSteel inserts properly then the castings would have to be complete.

4.5.2   The second run
4.5.2.1   Modifications
Based on the failure of the tool to produce complete castings from the first run, it was felt prudent to modify the tool to include a solid 'nitrided' central pin (a modified and fixed ejector pin) as can be found in the existing tool and shown in Fig. 4.13.

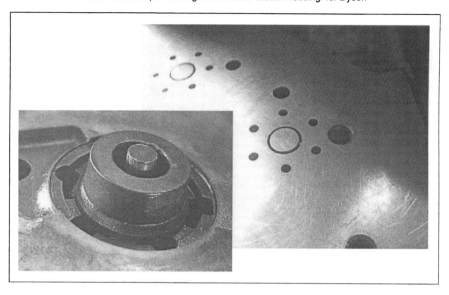

Fig. 4.13 Modified tool showing nitrided solid central pin feature

In addition to this work, the tool was polished further, based on the fears that the casting may have been adhering to a rougher surface in the first run. In fact, when the frozen part of the casting was removed from the tool it was noted that there was some pitting that was visible around the central core (this effect may have been formed by shot peening). This would suggest that surface finish is important as otherwise material on the surface of the tool may be pulled away by the action of the molten aluminium solidifying to it.

4.5.2.2   *Observations*
No changes were made to the operating conditions or machine parameters used for the run. As with the first run of 41 castings, high-temperature grease was used to ensure that the first few castings would be released.

Observations for this run were:

Casting 2:	Overflows froze in place on the 20 μm DirectSteel inserts as there were no ejectors in this part of the tool. It was decided to carry on and, in the end, there was no discernible effect on the castings.
Casting 45–50:	Castings were clean with good definition and clean ejection.
Casting 100:	Faint pitting visible on casting but no indication that the surface of the DMLS inserts was degrading. Clean castings, quality improving with each shot.
Casting 100–300:	Clean castings throughout this stage of the run. Castings good enough for production use (as stated by Dyson). A small area on the tips of two of the castellations froze in the tool.

Casting 350:	The frozen tip of the two castellations seemed to be casting out (it was reducing in size). Some pitting visible on the casting.
Casting 445–450:	Pull lines appearing on the casting where strata lines were on the castellations (see Fig. 4.12), run not expected to last much longer.
Casting 472:	Casting froze in tool as in first run shearing the casting at the central pin feature (see Fig. 4.11). Run terminated.

Including the castings from the first run, a total of 513 castings were produced which satisfied the objective for 500 castings. Due to time constraints, Kemlows could not remove the frozen casting without removing the tool to see how far the tool could be run on. Kemlows removed the frozen casting later but running the tool further ran outside the remit of the project. Figures 4.14–4.16 show details of the castings produced.

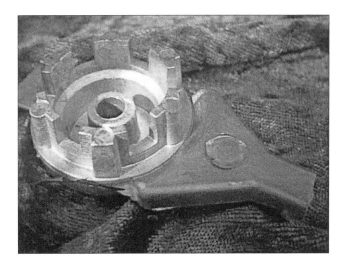

Fig. 4.14 Casting 146 front

Fig. 4.15 Casting 146 rear

Fig. 4.16 472 castings from second run

4.5.3   CMM comparison of castings against CAD model

From the samples taken during the second run CMM measurements were taken at the points shown in Figs 4.17 and 4.18.

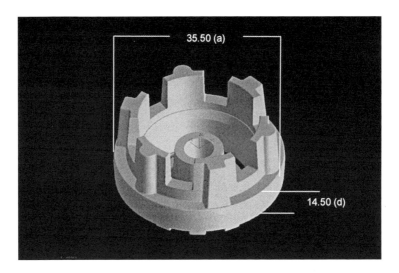

Fig. 4.17 Measurement points on casting for drift from tolerance

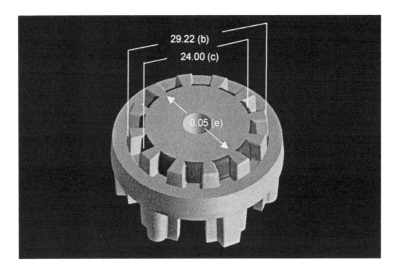

Fig. 4.18 Measurement points on casting for drift from tolerance

The results from the CMM analysis are shown in Table 4.2.

Rapid Tooling for Aluminium Die-cast Components – An Investigation into
EOS DMLS 20 µm Tooling Inserts for a 'Clutch Housing' for Dyson

59

Table 4.2 CMM report for casting tolerance

Measurement Type	Nominal	Upper tolerance	Lower tolerance	Sample number							
				98	149	198	249	298	348	397	446
a Dimension	35.50	0.15	-0.15	35.36	35.34	35.37	35.30	35.34	35.34	35.36	35.40
b Dimension	29.22	0.10	-0.10	29.25	29.23	29.18	29.19	29.23	29.24	29.18	29.24
c Dimension	24.00	0.15	-0.15	23.95	23.94	23.95	23.92	23.95	23.94	23.94	23.93
d Dimension	14.50	0.05	-0.10	14.10	14.13	13.97	14.01	14.11	14.12	14.16	14.14
e Flatness		0.05	0.00	0.02	0.07	0.09	0.02	0.03	0.03	0.03	0.04
f Parallelism		0.05	0.00	0.04	0.02	0.06	0.03	0.02	0.01	0.03	0.01
g Parallelism		0.05	0.00	0.02	0.01	0.02	0.01	0.01	0.01	0.01	0

Tolerances were set by Dyson dependant on dimension. Those data in italics are either above or below tolerance.

Eight samples were measured, one casting from each set of five taken from each fifty castings produced. Dimensions that were measured relate to dimensions 'a' through 'g' in Figs 4.17 and 4.18. Tolerances were defined from the engineering drawings originally supplied by Dysons to Kemlows.

What was encouraging is that most measurements lie within tolerance with the notable exception of 'a' and 'd'. Dimension 'a' is the largest of all the measurements so is more likely to drift over the 0.15 mm threshold, whereas dimension 'd' fell consistently outside tolerance and was probably an indication of either an error when building in the z-axis (which had been previously stated by EOS) or compaction through the effect of peening. The 'd' dimensions are not consistently under tolerance either but drift by 0.19 mm over the range, which could indicate distortion in the tool. Dimension 'd' is also interesting in that its dimension reduced over the run to the point at which the highest quality castings appeared (defined by visual inspection by Kemlows) and then increased towards the end of the run.

Correlating the smallest dimension on 'd' to the measurements taken for flatness 'e' and parallelism 'f' then a pattern emerges at castings 149 to 198 where three of the dimensions drift from tolerance. Again there is no clear reason but these castings were certainly the cleanest of the run and the tool was probably running at its optimum and this may have generated heat that could distort the inserts. Dyson's official summing up of the measurements was:

"The CMM results generally showed that the accuracy and repeatability of the components was acceptable. Although there were some dimensions out of tolerance, the consistency of the errors suggested that these were related to tool finishing issues, rather than any fundamental defects in the tool. In terms of the tool accuracy, most dimensions appeared to be within acceptable limits – this is difficult to quantify, as tool tolerances are not determined by Dyson. Generally, the process proved able to produce consistent and accurate parts to a high standard, which would certainly satisfy prototype and pre-production requirements."

4.6   Discussion

4.6.1   Observations on tool wear

As with the first run, during the second trial there was evidence of parts of the casting freezing in the tool and, ultimately, the casting froze on casting 472. At the termination of the second run it was unclear why the casting froze and it was suspected that the surface of the tool may have degraded to the point that castings could not be ejected effectively. On consideration and later investigation, this was not proved to be the case as casting 472 could have only shrunk on to the solid nitrided pin that made up the feature that caused the casting to shear (and not the tool). What was initially thought to be degradation of the tool surface was in fact the surface finish imparted by the shot peening process.

Up to this point, information relating to the finishing of the inserts by RPI Finland was not available and the group were unaware of the shot peening done and this may have led to the wrong assumption about tool degradation being made. Figures 4.19 and 4.20 show details of the tool surface and clearly show the rippled surface left by peening and not surface degradation as was thought during the trial. In fact the one area that should wear first on the tool would be the knife-edge feature of the inlet gate and this feature was still well defined showing no visible sign of excessive wear (shown in Fig. 4.21). Figure 4.22 shows the damage to the castellation feature that froze in the tool.

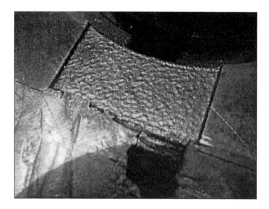

Fig. 4.19 Rippled effect left by shot peening on overflow channel

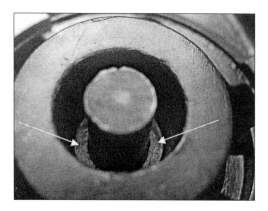

Fig. 4.20 Surface finish at base of hollow central feature (arrowed)

Fig. 4.21 Detail of knife-edge on inlet gate showing no visible sign of wear (circled)

Fig. 4.22 Castellation damage on casting 146 by partial freezing of casting in tool

4.6.2   Observations on casting porosity

A second issue that arose with the run was that of porosity in the castings. It was noted previously that during the first few castings of the second run, the overflows froze in place. This effectively blocked the overflows to each cavity but was not sufficient to stop the run. It was expected that porosity and impurity would be seen early in the casting run but this proved not to be the case.

By casting 149 the parts coming from the 20 μm DirectSteel inserts were almost faultless (in Dyson's view) and it was speculated that trapped gas may be still venting somewhere as material enters each cavity or again inclusions and voids would be seen in the casting. From the outset of the run it was noted that the 20 μm insert had a matt appearance compared to the conventional insert and this could possibly be explained either as the finish imparted through shot peening or possibly due to a small amount of porosity (potentially less than one per cent)

in the 20 µm DirectSteel inserts. It was, therefore, speculated whether a small amount of porosity would be sufficient to vent the die cavity as molten aluminium entered the die.
This effect did not last for long and by casting 248 onwards small pockets of trapped gas began to pierce the surface of the casting as shown in Fig. 4.23 (small surface defects are also visible on this casting).

Fig. 4.23 Porosity on casting 248 (circled)

4.6.3   Design implications of 20 µm DirectSteel tooling

The use of the 20 µm DirectSteel central pin for the first run confirmed that in this situation a solid pin would be required. However, the group initially assumed that this section would need to be replaced because the pin would simply shear off on the first casting. This assumption was proved wrong. The first run was terminated as the entire central core feature froze in place and it was found that upon ejection it was the casting that sheared not the tool surface that broke away.

From experience with die-casting with 50 µm DirectSteel tools this is a notable improvement in the mechanical properties of the new material. Shearing across grain boundaries/layers thwarted previous attempts to die cast with the 50 µm DirectSteel material.

Polishing is an issue with the 20 µm DirectSteel material. As to the degree of polishing required this research was not able to quantify. Even so, the surface of the inserts remained in tact and resisted soldering and thermal fatigue.

Of initial concern were the cooling lines and sealing with epoxy resin. However, not only did the lines remain sealed throughout the run but showed no indication of any reduced thermal efficiency due to the presence of the resin within the tool. In addition, the inserts were initially heated using the cooling as heating lines. Hot oil up to 500 °C was used and again this led to no degradation in the effectiveness of the cooling lines. Conformal cooling/heating channels in these tools is currently under investigation.

4.6.4   Costings for the parts produced

The time taken to construct the existing Dyson tool was ten weeks with a further two weeks then required for sampling. Time to build the DMLS 20 μm Direct Steel inserts (male and female) was 179 h (83 h + 96 h) plus 48 h finishing, 8 h set-up, giving 235 h.

It should be noted that the 179 h taken to build the inserts was deemed excessive by EOS as the inserts were larger than would normally be built by the DMLS process. EOS estimated that a more realistic time would have been 80–100 h had the design been optimized so that both male and female inserts fitted on to one build platform and their sizes reduced. To have achieved this, however, would have required a secondary clamping system to be built for the DMLS inserts in order to get them into the Dyson bolster and it was dubious whether this would have saved time overall. In reality, if designing a tool from scratch, changes would be made to both die and bolster (but again the design must not drift too far from the production tooling design if it is this that is being appraised).

The cost of a prototype die-cast tool is never going to produce castings cheaper than the production tool. The cost to produce the existing Dyson tool was £22 250 (£2250 of this included the trimming tool), which related to a cost per casting of approximately £0.13 over its life. The estimated cost (EOS did not charge for the experimental work) for the production of the inserts would have been €10 000–15 000. Cost per casting would normally include the number of castings divided into the cost of the tool (plus the materials cost) but in this case, this was complicated as no more castings were produced after casting 513 when, in hindsight, the tool may have produced many more castings. Cost per casting for this project was €19–29 per casting (or €10 per part) which relates to £12 per casting or £6 per part (two parts on each casting) for the non-optimized inserts. If the run had been extended and an optimized strategy employed then this figure may have been closer to €3–6 per part (€5000–8000/700 castings/2 parts).

Conclusions

- The DMLS 20 μm DirectSteel system has been proved to withstand high-pressure die-casting with aluminium LM24 under full production conditions for 513 shots.
- The tool began to show signs of imminent freezing after 450 castings (draw lines appeared on the castellations).
- No signs of hot cracking (heat checking) were evident on the inserts after 513 castings.
- From visual inspection during the second run the inserts ran at their optimum between castings 149–198 but during this phase the tool also produced signs of minor distortion.
- Similar issues apply when designing tall narrow features into a 20 μm DirectSteel tool as apply for production tool, i.e. isolated pin sections will cause problems.

- The inclusion of heating/cooling lines within the inserts was thought to be a problem but proved to work satisfactorily during the trials. The use of an epoxy resin infiltrant around the cooling lines also proved sufficient to withstand pressurized water/hot oil and was still able to remove heat from the tool effectively during the casting cycle.

Dyson's summary report states:

"The DMLS process has proven itself to be a viable tool for the production of limited batches of pressure die-castings in aluminium. This is very encouraging – if the process can cope with die casting, it will certainly be capable of handling substantial volumes of most plastic parts, including more demanding applications such as thermosets and filled materials. The only limitation to the process currently appears to be finishing of the tools; however, the project has proved very worthwhile for us as a technology demonstrator."

Acknowledgements

The author would like to thank the project partners and key people from those organizations that contributed to, and made possible, this project, namely:

Tatu Syvanen   *(EOS Finland)*
Olli Nyrhila   *(EOS Finland)*
Roger Dunlop *(Dyson Limited)*
Chris Lamont *(Kemlows Die-casting)*
Paul Phillips   *(Kemlows Die-casting)*

R Soar
Rapid Manufacturing Group, Loughborough University, UK

5

Rapid Route for Investment Casting

C Ryall, S Zhang, and D Wimpenny

Abstract

This Chapter presents the results of a recent study into the quality of castings produced from a wide range of rapid prototype patterns. The study, based on real aerospace components, compared six alternative rapid prototyping routes on the basis of accuracy, surface quality, and ease of casting, including casting integrity by x-ray analysis to aerospace standards.

5.1 Introduction

Casting, in particular investment casting, is one of the key application areas for rapid prototyping. The majority of rapid prototyping methods are capable of producing models that can be used as lost patterns for investment casting. However, the level of success can vary significantly between the different RP processes. A number of research programmes have been conducted to develop and compare the use of RP models as lost patterns for investment casting (1-4). Although these studies have led to some important developments in this field, the results become quickly outdated due to evolution of both the RP equipment and materials.

Investment casting of RP models could play a significant role in the aerospace industry, however, there has been reluctance to apply this technology. To be more widely adopted the aerospace companies must be able to show that the process can consistently produce aircraft components that meet the rigorous standard required for flight certification.

5.2 IMI RAMAC programme

In February 1999, IMI RAMAC (**Ra**pid **M**anufacture of High Integrity **A**erospace **C**omponents) programme commenced. Under the programme a broad range of manufacturing methods, that can be used to generate high integrity metal parts, are being investigated and compared. These methods include high-speed machining, casting from rapid prototype models, and rapid tooling to produce conventional wax patterns for investment casting. The programme also includes an evaluation of reverse engineering, not only for gathering data from existing components, where CAD data is not available, but also as an inspection method for measuring the geometry of components. The information generated from these trials will be fed into an expert system that will assist engineers to select the most appropriate manufacturing route for a particular part.

The aim is to provide the aerospace industry with a cost effective manufacturing process for high integrity metallic components, in the production volumes required for the aerospace market. The industrial partners in the programme include aerospace end-users (TRW Lucas Aerospace, GKN Westland Helicopters, BAE Systems), an investment casting foundry (Aeromet International), computer hardware and software developers (Sun Microsystems and SDRC), and a CNC machine tool manufacturer (Bridgeport Machine Tools Limited).

This Chapter presents the results of the initial trial, which assessed a range of RP techniques for the manufacture of 'lost' patterns for investment casting of aerospace quality castings.

5.3   Programme of work – investment cast oil jets

The aim of this initial trial was to evaluate a wide range of RP methods for the manufacture of wax patterns for investment casting. The results of this trial will be used to select the most suitable RP routes for investment casting. These routes will then be developed further and scaled-up using larger, more complex aerospace parts.

The results of the trials were evaluated on the basis of:

- accuracy;
- surface finish;
- visual inspection of RP patterns and final casting;
- compatibility with the standard casting process;
- casting quality based on x-ray analysis.

To limit the cost and time to manufacture the RP models and produce the final castings a small, relatively simple part was selected for these trials. This component, the GKN Wesland oil jet, is shown in Fig. 5.1.

Fig. 5.1 Sanders model of the GKN Westland oil jet

The RP methods used for the trial are detailed in Table 5.1, below.

Table 5.1 List of RP models used in trial

Technology	Machine	Material and model form
Stereolithography	SLA5000	SL5195 QuickCast II
Selective Laser Sintering	DTM Sinterstation 2500	DTM CastForm DTM TrueForm
Laser sintering	EOSint P350	EOS Polystyrol
3Dsystems MJM	Actua 2100	Thermojet 75
Sanders 3D Plotting	Model Maker II	ProtoBuild
Stratasys FDM	FDM1650	ABS P400

5.3.1 Test procedure
In order to check the consistency of the different RP systems it was decided to produce six models by each of the different RP routes. The test procedure applied was:

- manufacture RP models;
- geometric inspection of models;
- visually inspect models and record any anomalies;
- investment cast models;
- subject casting to x-ray inspection.

5.3.2 RP model manufacture
The models were constructed using the build conditions shown in Table 5.2 below.

Table 5.2 Build details for the RP models

Model type	Build time for 6 models (hours)	Layer thickness (mm)	Cost	Models manufactured by
EOS Polystyrol	8.6 [e]	0.15	As Trueform	EOS
DTM TrueForm	5.5 [e]	0.1	1 off £189 6 off £231	Rover Group
DTM CastForm	5.5 [e]	0.10	No details available	DTM
Sanders	50	0.05	1 off £200 6 off £1200	Alpha Mechanical Engineering Limited
Actua	6.5 [e]	0.05	1 off £40 6 off £66	Rover Group
FDM	15 (solid build)	0.25	No details available	Laser Lines Limited
SLA QuickCast	3	0.15	1 off £209 6 off £360	Rover Group

[e] estimated time

It can be seen that the longest build time is for the Sanders machine, which took almost eight times longer than the Actua machine to build models with the same, 50 micron, layer thickness. The Sanders models were purchased from an external bureau (the University has since purchased a Sanders machine) and it is notable that the models cost £200 each, irrespective of whether one or six are produced. The SLA5000 built the models in the shortest time, however, the Actua models were the lowest cost to produce at only £11 each, if six are manufactured together. The build time of 15 hours for the FDM models could have been reduced if they had been produced with a hollow core.

5.3.3   Geometric inspection of RP models

The inspection procedure was based on the methods used to evaluate the current production components. The following items were checked.

- Oil jet orientation with respect to the three location holes (Fig. 5.2).
- Angle of the oil jet pipe (Fig. 5.3).
- Diameter of the jet pipe at two positions (near the exit point and the base).
- Roundness of the oil pipe.

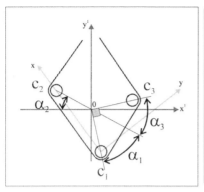

**Fig. 5.2 Oil jet orientation with respect to the three location holes (left)
Sanders model (right)**

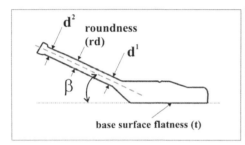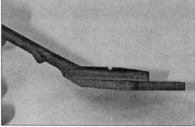

Fig. 5.3 Angle and diameter of oil jet pipe (left) Sanders model (right)

A Renishaw cylone was used to measure the RP models (Fig. 5.4). This is a contact scanning system, which is generally used for reverse engineering to digitize an 'unknown surface'. In this case the Renishaw cyclone, together with its associated software, Truecut, was used to measure the entire surface of the oil jets, rather than gather point data using a conventional CMC fitted with a touch trigger probe. To completely measure the oil jets, three separate scan operations were required. First, the upper surface of the oil jet was scanned, including the top half of the oil jet pipe. Second, the underside of the oil jet was scanned, by securing the models on a simple fixture. Third, the diameter of the oil jet pipe was measured precisely, by orientating it vertically on the bed on the machine. The point cloud files generated were processed using Surfacer 8 software and combined to give the surface information for the entire model.

Fig. 5.4 Renishaw cyclone

The results of the geometeric inspection of the RP models are shown in Table 5.3 below.

Table 5.3 Geometric inspection data for RP models

	Oil jet orientation			Angle oil jet pipe	Diameter of pipe		Roundness of jet pipe	Flatness of base
Parameter	α_1	α_2	α_3	β	d^1	d^2	rd	t
Theoretical values	45°	37°	45°	27°	6.04 (mm)#	6.04 (mm)#	0 (mm)	0 (mm)
EOS Polystyrol	45.22	38.09	45.21	27.22	6.35	6.18	0.24	0.37
DTM TrueForm	45.25	37.60	44.67	26.69	6.00	5.80	0.18	0.40
DTM CastForm	43.39	38.43	46.81	24.72	6.04	5.78	0.18	0.35
Sanders	44.98	37.87	45.25	27.89	6.04	5.76	0.15	0.39
Actua	44.87	37.76	45.00	27.46	6.32	5.98	0.22	0.40
FDM	45.03	37.72	45.03	27.69	6.06	5.87	0.23	0.66
SLA QuickCast	45.22	38.09	45.21	27.21	6.34	6.17	0.23	0.37

Including casting allowance

The DTM CastForm models showed some significant errors, for example, the angle of the oil jet pipe and the orientation of the holes. However, this could be due to distortion caused during the wax infiltration of the models. In addition, this is a recently introduced material and it is likely that the accuracy of models will improve with increasing experience. The diameter of the oil jet pipe for the SLA models appeared oversized, however, this could be due to distortion of the pipe caused by the location of the supports. The pipe diameter of the EOS Polystyrol models was also oversized, but this could be due to a build up of the resin used to infiltrate the surface. The accuracy of the Sanders models, though this is reputed to be the most precise RP systems, was generally no better then either the FDM or Actua models. The values for the flatness of the base must be regarded with some caution as, in many cases, the base has been artificially flattened by abrading.

5.3.4   Visual inspection of models

Each of the model types was examined and visually assessed on the basis of:

- surface finish;
- feature definition;
- quality of holes in the base.

The results of this assessment are shown in Table 5.4 below.

Table 5.4 Results of visual inspection of the RP models

Model type	Visual assessment
EOS Polystyrol	Surface has granular texture Features rather indistinct Holes not perfectly round
DTM TrueForm	Surface quite smooth Good feature definition Holes round
DTM CastForm	Surface smooth Feature not a clearly defined as TrueForm models Holes are slightly oval
Sanders	Very good surface finish Best feature definition Holes perfectly round
Actua	Very smooth upward facing surface Good feature definition on upward facing surfaces Holes round
FDM	Surface shows patterns of filament Features not a well defined as Sanders or DTM models Holes slightly irregular
SLA QuickCast	Smooth upward facing surfaces but support marks showing on lower surfaces Very good feature definition Holes round

Of the different systems the Sanders models gave the best combination of surface finish, feature definition, and hole quality. Both the Stereolithography and Actua models gave excellent surface quality and reproduction of features, but were marred by the witness of supports on the underside of the models. This problem could have been reduced by additional clean-up operations. In the case of the Actua models, recent improvements in support removal techniques include the use of paraffin to manually polish the underside of models. However, there were concerns that these additional processes would undermine the accuracy of the parts and for this reason it was decided to employ the minimum amount of manual post-processing on the models. Although the quality of the FDM models was generally very good, with no witness from the supports, there was a pattern on the surface of the models produced by the extruded tracks. Of the laser sintered samples the DTM Trueform models had the best surface finish, feature definition, and hole quality. TrueForm has been superseded by CastForm for the manufacture of investment casting patterns. Unfortunately, although quality of the Castform models was good, they lacked the sharp feature definition of the Trueform models. The EOS Polystyrol models were rather disappointing, having a granular, gritty surface with very poor feature definition.

5.3.5   Investment casting

Only three samples from each system were processed into castings, in order to retain models from each RP system for future analysis. Where possible the standard investment casting procedure was used, as described below.

- *Assembly* – Three samples from each modelling system were assembled together on a single wax runner bar attached to a standard wax cup.
- *Shelling* – The assemblies were given six coats of ceramic and dried using the standard production method.
- *Autoclave* – The ceramic moulds were de-waxed using 7bar steam pressure for a duration of 12 minutes.
- *Firing* – Ceramic moulds were fired at a temperature between 750–900 °C for a period of 15–45 minutes.
- *Shakeout* – Ceramic moulds were inverted and shaken to remove any ash or ceramic debris.
- *Casting* – Ceramic moulds were preheated to a temperature between 400–500 °C, prior to casting aluminium L65 at a temperature of 700–800 °C.

Ideally, the RP models should be processed in exactly the same way as conventional wax patterns. This is not only for reasons of convenience for the foundry; by employing the standard foundry route this avoids the need to re-qualify the process in order to gain flight certification for the castings. Unfortunately, although it was possible to use the standard approach with the Actua, Sanders, and laser sintered models, the process had to be modified for the FDM and Stereolithography models.

ABS patterns produced by the FDM process are not as widely used for investment casting as other routes, such as the Stereolithography for example. Because of the lack of knowledge in the field, Stratasys have produced guidelines based on the result of trials at six US foundries (5). These trials showed that it is possible to investment cast ABS patterns, but the process must be modified accordingly. The ABS patterns are unsuitable for the autoclave process and the shell must be placed into a flash furnace, reaching a temperature of over 1000 °C, which reduces the pattern to ash. A more detailed procedure is described in Table 5.5 below.

Table 5.5 Recommended investment casting condition for use with FDM ABS patterns (5)

Preheat and load temperature	Ramp to maximum temperature	Holding time	Cooling
871 °C	1066–1120 °C	1–2 hours	Overnight

QuickCast II, with its hexagonal internal structure, was developed to enable models to be autoclaved without shell cracking, thus enabling foundries to use QuickCast models in the same way as conventional wax patterns. Unfortunately, initial trials on the oil jet resulted in shell cracking during the autoclave process (this has been reported by other foundries) and as a consequence it was decided to flash fire all subsequent QuickCast II patterns.

The results of the casting trial are shown in Tables 6–12, below (data courtesy of Aeromet International).

Table 5.6 Cast trial results/observations for Sanders models

Process stage	Results/observations
Autoclave	Produced good moulds no cracks found
Firing	No deposit found in the mould
Shakeout	No loose deposit found in the mould
Casting	Good castings produced
Penetrant flaw detection (PFD)	Passed PFD no surface defects found
X-ray analysis	Passed [1]

Table 5.7 Cast trial results/observations for Actua models

Process stage	Results/observations
Autoclave	Produced good moulds no cracks found
Firing	No deposit found in the mould
Shakeout	No loose deposit found in the mould
Casting	Good castings produced
Penetrant flaw detection (PFD)	Passed PFD no surface defects found
X-ray analysis	Passed [1]

Table 5.8 Cast trial results/observations for TrueForm models

Process stage	Results/observations
Autoclave	Model left in mould no cracks found
Firing	Black ash deposits found in the mould
Shakeout	No loose deposit found in the mould
Casting	Good castings produced
Penetrant flaw detection (PFD)	Passed PFD no surface defects found
X-ray analysis	Passed [1]

Table 5.9 Cast trial results/observations for CastForm models

Process stage	Results/observations
Autoclave	Part of model in mould but no cracks found
Firing	No deposit found in the mould
Shakeout	No loose deposit found in the mould
Casting	Good castings produced
Penetrant flaw detection (PFD)	Passed PFD no surface defects found
X-ray analysis	Passed [1]

Table 5.10 Cast trial results/observations for EOS Polystyrol models

Process stage	Results/observations
Autoclave	Produced good moulds no cracks found
Firing	No deposit found in the mould
Shakeout	No loose deposit found in the mould
Casting	Moulds split during casting – no parts produced
Penetrant flaw detection (PFD)	–
X-ray Analysis	–

Table 5.11 Cast trial results/observations for QuickCast models

Process stage	Results/observations
Autoclave	–
Firing	Flash fired
Shakeout	Powder residue found in mould
Casting	Good castings produced
Penetrant flaw detection (PFD)	Passed PFD no surface defects found
X-ray analysis	Passed [1]

Table 5.12 Cast trial results/observations for FDM ABS models

Process stage	Results/observations
Autoclave	–
Firing	No deposit found in the mould
Shakeout	No loose deposit found in the mould
Casting	Good castings produced
Penetrant flaw detection (PFD)	Passed PFD no surface defects found
X-ray analysis	Passed [1]

[1] Porosity level 1: ASTM E155. (Castings considered acceptable to GKN Wesrland Helicoptor Standard: WHPS 648 High stress areas and all over other areas.)

The Actua and Sanders models were the easiest to process, giving no residue after firing and producing good quality castings. The TrueForm model produced noticeable levels of ash after firing. This problem has been overcome with the CastForm models, which did not leave any residue after firing. All the RP routes produced sound castings, apart from the EOS Polystyrol patterns where failure of the shell occurred during pouring if the metal. There is always a risk of shell failure, even with conventional wax patterns, and it is possible that some cracking had occurred during firing which had gone unnoticed. It is notable that some ash residue was present after the QuickCast models had been flash fired.

5.3.6 Visual inspection of the castings
To assess whether the surface quality and feature definition of the RP models had a noticeable influence on the quality of the final castings produced, they were examined and visually assessed. The results of this assessment are shown in Table 5.13 below.

Table 5.13 Visual assessment of castings produced

Model type	Visual assessment
EOS Polystyrol	No castings produced
DTM Trueform	Surfaces have a slightly granular texture which seems rougher than on the original RP models
DTM Castform	Surface smoother than TrueForm, with no evidence of stair steeping, but some loss of feature definition
Sanders	Casting surface smooth with excellent feature definition comparable with castings from conventional wax patterns
Actua	Upper surfaces comparable with the Sanders model but underside showing evidence of the supports
FDM	Very little evidence of supports, however, there is a witness of the start/finish of the extruded tracks on the pipe and some stair stepping.
SLA	Stair stepping clearly visible on some of the upward facing surfaces. Evidence of the supports on the underside of the pipe.

In most cases the quality of the casting closely reflected the quality of the RP patterns used to produce them. Undoubtedly, the best surface finish and feature definition was produce from the Sander models. Indeed, the castings were comparable with those produced from conventional wax patterns. The upward facing surface of the castings produced from the Actua models were almost as good as those from the Sanders patterns, but the underside showed a poor surface finish due to the supports. The CastForm models produced a casting with a smoother surface than the TrueForm models, with no visible stair stepping, but there was a significant loss of feature definition. The castings from the SLA models showed stair stepping on the upward facing surface and distortion on the underside of the pipe, due to the supports. The castings from the FDM models faithfully reproduced many of the flaws on the models, including the witness of the extruded tracks (just visible) and the start and finish of the tracks, particularly on the pipe.

Conclusions and further work

All of the RP routes, with the exception of the EOS Polystyrol patterns, produced good quality castings, which meet the x-ray standards set by the customer (GKN Westlands). In the case of the EOS Polystyrol patterns, the shell failed during casting and further trials must be performed. It is probable that the shell failure was the result of the relative inexperience of the foundry (this is their first attempt at casting this type of RP model). Other foundries around Europe, particularly in Germany, successfully cast using EOS Polystyrol models on a regular basis.

It was very difficult to deduce any firm conclusions from the geometric inspection of the RP models, particularly in light of the small number of samples measured. However, none of the samples showed errors that would give grave cause for concern, with the exception of the CastForm samples. However, it is likely that the accuracy of CastForm model will improve as more experience of this recently introduced material is gained. All of the other RP models were sufficiently accurate to meet the needs of investment casting patterns. Where higher accuracy is required it is common practice to add a machining allowance to the castings, and many features on the final casting, such as the angle of the pipe, would normally be manually adjusted to comply with the required tolerance.

The visual examination of the RP models showed that the Sanders machine produced the best surface quality and feature definition of all the systems evaluated. The quality of the models was reflected in the castings produced, which were comparable with those from conventional wax patterns.

Based on results to date for small castings, the most appropriate RP routes appears to be the Actua and Sanders systems. These machines give 'wax' patterns that are completely compatible with the standard investment casting process, and have an excellent combination of accuracy and surface finish. Unfortunately, the excruciatingly slow build speed of the Sanders machine is its Achilles heel. For low volume production the Actua machine is a more practical solution, provided that the relatively poor surface quality of the underside of the models is acceptable.

Further work on the oil jet is currently being performed. Thermojet models have recently been manufactured and these are currently being converted to castings. Initial observations indicate that there is an improvement in the quality of the upper and lower facing surfaces, coupled with a reduction in build time. Over the next two years, the investment casting trials will extend to larger more complex castings to provide a comprehensive set of data that will form the basis of the process selection expert system.

Acknowledgements

The authors of this Chapter extend their gratitude to EPSRC and the industrial partners for their support for the RAMAC programme.

References

(1) **Dickens, P. M.** *et al.* Conversion of RP Models into Investment Castings, Rapid Prototyping Journal, Vol. 1, No. 4, 1995, MCB University Press, pp 4–11.

(2) **Wimpenny, D. I.** Rapid Prototyping: Part of a Computer Integrated Approach to Investment Casting, 23rd European Conference on Investment Casting, Prague, June 1996, Paper 14.

(3) **Hague, R.** and **Dickens, P. M.** Finite Element Analysis Calculated Stress in Investment Castings Shells using Stereolithography Models as Patterns, 5th European Conference on Rapid Prototyping and Manufacturing, 1996, pp 31–46.

(4) **Koch, M.** Rapid Prototyping and Casting, 3rd European Conference on Rapid Prototyping and Manufacturing, 1994, pp 73–76.

(5) **Gouldsen, C.** and **Blake, P.** Investment Casting Using FDM/ABS Rapid Prototype Patterns, June 1998.

C Ryall, S Zhang, and **D Wimpenny**
Warwick Manufacturing Group, University of Warwick, Coventry, UK

6

The Use of QuickCast™ and Investment Casting for the Production of Novel A-posts for a Functional Safety Concept Car

N Hopkinson and **J Almgren**

This Chapter describes a project performed by De Montfort University's Rapid Manufacturing Consortium along with Volvo Cars and Land Rover. The project was performed to find a suitable manufacturing method to produce novel A-posts for a one off functional car. The chosen method of investment casting from a stereolithography pattern was proved using a test section of one of the final parts to be made. Finally, the complete castings for both left and right hand A-posts were created and fitted to the car which has been used for extensive test-driving.

The authors assume that the reader is aware of both the investment casting process and the principle rapid prototyping (RP) technologies available today.

6.1 Introduction

In early 2001 at the Detroit Autoshow, Volvo Cars, who are part of Ford's Premier Automotive Group (PAG), unveiled a new 'Safety Concept Car' (SCC) which is shown in Fig. 6.1.

Fig. 6.1 The Volvo safety concept car

The SCC includes a novel A-post design that involves a lattice of glass panels that are intended to increase the available forward visibility for the car occupants. Figure 6.2 shows views from inside and outside of the car, highlighting how the A-post incorporates voids that have been designed to allow a more complete view from the driving position.

Fig. 6.2 Views through the A-post

At the start of this project, Volvo had already produced a non-functional concept car shown in Fig. 6.1. The A-posts in the non-functional car were produced in nylon by laser sintering using an EOSP 360 machine. However, the next goal was to produce a functional version that could be used for live tests and presented to the world's media – clearly metal versions were required.

Traditionally, A-posts are manufactured by sheet metal stamping. However, the unusually complex design, and the fact that only one functional concept car was envisaged, dictated that traditional methods of manufacture could not be cost-effectively employed due to the prohibitive lead times and costs of tooling. Therefore, in order to include the A-post design in the functional SCC a different method of manufacture was needed to enable it to be safely operated and tested. One approach which had been considered by Volvo was to machine the A-posts from billet steel or aluminium. As a member of De Montfort University's Rapid Manufacturing Consortium (RMC), Volvo Cars approached the RMC to see if there was an alternative solution for the production of the A-posts. Having examined CAD models of the parts, it was suggested that an ideal route would be to create stereolithography models built in the QuickCast™ build style and then use these models as sacrificial patterns in the investment casting process. The parts could then be cast in the material of Volvo's choice. Staff within the RMC have much experience of QuickCast™ investment casting and have performed similar technology transfer projects for other RMC members. Additionally, it was suggested that Volvo contact Graham Tromans of Land Rover (also part of the PAG) who has considerable experience in this field.

In order for Volvo to be confident that this approach would be successful, it was suggested that a test casting should be produced to prove the concept of manufacture. Successful production of the test part would likely lead to the production of the full A-posts by the investment casting process.

6.2   Project goals

The main goal of this project was to produce the full, left and right, A-posts to be used in the functional concept car. This part of the project was subject to the initial part (i.e. producing a test casting) being successful.

An additional goal of the project concerned technology transfer and was to demonstrate to key Volvo personnel the potential for producing one-off (or short production run) metal castings using QuickCast™ stereolithography parts as sacrificial models in the investment casting process. This would allow Volvo staff to use the technique, for the production of special A-posts, for the new functional safety concept car and, if applicable, for future projects.

6.3   RP and investment casting

The initial evolution of RP, in the early 1990s, saw a significant amount of work with investment casting. The two technologies were well suited given that RP provided a quick means of producing a master pattern and that the material properties of RP parts were not good enough for functional use. Indeed, the poor mechanical properties of RP parts proved an asset to investment casting due to the ease with which they could be burned out from the casting shell.

Stereolithography was the first RP process to be used in earnest for investment casting (1), with early problems of pattern expansion during burnout being remedied by the development of quasi-hollow build structures, such as QuickCast™ (2, 3). Numerous examples of the use of QuickCast for investment casting have been published (4).

Probably spurred on by the early successes of RP with investment casting, other RP processes were developed which were suited to investment casting. The Sanders Model Maker was aimed at producing fine, detailed, small patterns for investment casting particularly in the jewellery industry. 3D Systems Actua and latterly ThermoJet 3D wax printers, produced wax parts which are traditionally associated with investment casting (5). Fused deposition modelling (FDM) from Stratasys also provided an option to produce wax patterns. It was also found that ABS patterns by FDM could easily be used for investment casting (5).

Powder-based processes including laser sintering from EOS GmbH and DTM Corp could be used to produce investment casting patterns in polystyrene **(5)**. The three-dimensional printing process developed at MIT and commercialized by Z-Corp also proved a means of producing investment casting patterns by RP **(5)**. The three-dimensional printing process also provided a solution for producing not the master, but the shell, by directly printing hollow ceramic parts for subsequent firing and casting, this was commercialized by Soligen in the USA.

6.4   Investment casting with QuickCast™ patterns

The stereolithography (SL) process created three-dimensional objects in photocurable epoxy resin using a scanning laser to selectively cure and solidify a liquid polymer in successive layers. SL parts may be built as solid parts (using the ACES™ build style) or as quasi-hollow parts with an internal honeycomb structure known as QuickCast™.

The QuickCast™ buildstyle was developed so that SL parts could be created as investment casting patterns, which would collapse in on themselves during burnout and reduce the possibility of the ceramic shell cracking under thermally induced stresses from an expanding pattern. QuickCast™ parts are not fully solid, they use significantly less material and require less laser scanning time than their solid counterparts, however, uncured liquid resin which remains inside a part during and after building must be drained prior to subsequent processing. Draining parts can prove problematic, especially for large parts and thin sections, where surface tension in the relatively viscous liquid resists gravitational or centrifugal forces which may be used to drain excess fluid.

SL parts require some manual finishing after building to remove supports, this is followed by a post curing procedure in which parts are subject to ultraviolet light which cures any excess liquid both outside and, especially in the case of QuickCast™, inside a part. When a QuickCast™ part is to be used for investment casting, it will require some finishing work such as smoothing of rough surfaces and sharp edges and filling of holes, such as those used to drain excess liquid from inside the structure.

6.5   Discussion of the work performed

Within the project there were several phases into which the work was divided. The project phases were as follows.

Phase 1   Examine CAD file of A-post to confirm that the most suitable method for manufacture has been chosen.
Phase 2   Receive production criteria from Volvo.
Phase 3   Select part of the A-post to be used as a test for manufacturability and suitability.
Phase 4   Produce QuickCast™ master pattern of test part.
Phase 5   Produce casting of test part.
Phase 6   Analyse finished test part for suitability.
Phase 7   Produce QuickCast™ master patterns of full A-posts.
Phase 8   Produce castings of full A-posts.
Phase 9   Finishing work to A-posts and assembly to the car.
Phase 10   Analysis of results.

The following is a discussion of the work performed during the project. For ease, the discussions focus on the various phases of the project.

6.5.1   Phase 1 – Examine CAD file of A-post to confirm that the most suitable method for manufacture has been chosen

Volvo sent a CAD file (in STL format) to the RMC for evaluation to ensure that investment casting using a QuickCast™ master pattern would be the most appropriate method of manufacture. The initial decision to use this manufacturing route was based on two-dimensional drawings seen at Volvo's headquarters in Gothenburg. A 1/5 scale physical model was built using fused deposition modelling to help evaluate the design for manufacturability (see Fig. 6.3).

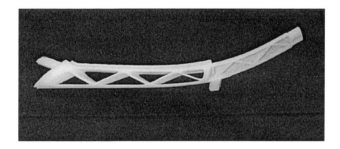

Fig. 6.3 Picture of 1/5 scale model built by FDM to evaluate manufacturability

6.5.2   Phase 2 – Receive production criteria from Volvo

Volvo supplied the production requirements of the part to ensure that the required materials could be used for investment casting. The requirements given by Volvo were initially quite loose, as there was no benchmark or existing product for comparison, and were described as follows.

- Material choice – aluminium or steel (to be decided after first test casting).
- Dimensional tolerances required – also to be decided after the first test casting.
- Surface finish – nice and sharp as the car is a functional show car.

6.5.3   Phase 3 – Select part of the A-post to be used as a test for manufacturability and suitability

Rather than building the complete A-posts, it was decided to build a test-section that may be evaluated for ease of manufacturing. The test section selected, shown in Fig. 6.4, reflects all the design intent of the actual A-posts required. In particular the test casting included thin sections, which might have proven difficult to fill effectively with molten aluminium or steel. It was felt that the geometry would not be difficult to produce by SL.

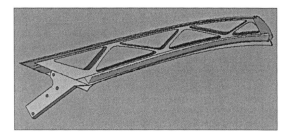

Fig. 6.4 STL file for the test section of the A-post

6.5.4   Phase 4 – Produce QuickCast™ master pattern of test part

A QuickCast™ sacrificial pattern for investment casting was produced by Land Rover. The optimum orientation of the part was chosen to give the best surface finish and to facilitate drainage. After draining, the pattern was post-cured in an ultraviolet oven, to cure any uncured resin, and sent to Tritech Precision Products for casting. Figure 6.5 shows the QuickCast™ master pattern of the test part.

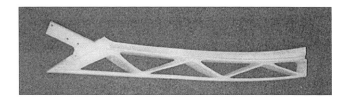

Fig. 6.5 QuickCast™ master pattern of the test part

6.5.5   Phase 5 – Produce casting of test part

Tritech Precision Products performed investment casting using aluminium (LM25TF) as selected by Volvo. This process took three weeks and has been broken down as follows.

Step 1	2 days	Dimensionally check and check QuickCast™ model (small radii added where the process dictates).
Step 2	1 day	Assembly on to sprue.
Step 3	7 days	Apply shell – 7 days for aluminium (14 days for steel).
Step 4	1 day	Burnout pattern.
Step 5	1 day	Cast.
Step 6	1 day	Remove from shell, cut off sprues, and clean-up.
Step 7	2 days	Heat treat for correction.

6.5.6   Phase 6 – Analyse finished test part for suitability

Figure 6.6 shows the test casting produced. This was sent to Volvo where it was assessed using the criteria laid out by Volvo in Phase 2. Volvo performed a geometric and material analysis of the test casting, to see if aluminium was suitable to use for the final A-post geometry. The casting was compared with other manufacturing routes, including milling. Volvo concluded that investment casting using SL QuickCast™ patterns was the best available route for producing the full A-posts, and that steel (alloy: ANC3B) should be used in the final A-posts.

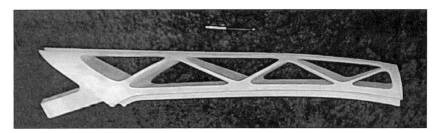

Fig. 6.6 Test casting produced in aluminium

6.5.7   Phase 7 – Produce QuickCast™ master patterns of full A-posts

Having ascertained the suitability of investment casting with QuickCast™ patterns, it was important to monitor times and costs for producing the full castings, so that comparisons with machining could be made. The STL files for the full A-posts were sent to Land Rover so that QuickCast™ patterns could be created, as described in Phase 4. Figure 6.7 shows one of the full patterns produced. Details of the time required to produce the patterns are given in Table 6.1.

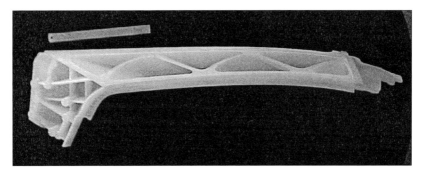

Fig. 6.7 One of the full patterns produced

Table 6.1 Times required to produce patterns for full A-posts

Operation	Time
Orientation and support generation in SL Software	4 hours
Slicing to create build file	32 minutes
Build time on SL5000	60 hours
Draining honeycomb structure	4 hours
Clean-Up Time (this includes support removal, final draining, bonding and finishing)	2 Days

Cost to produce – SEK39000 (£2613) for one A-post
                   SEK42000 (£2814) for the other A-post

6.5.8   Phase 8 – Produce castings of full A-posts

Land Rover produced the QuickCast^{TM} patterns and then sent them on to Tritech Precision Products for investment casting. The investment casting procedure, described in Phase 5, was used with a total lead-time of 30 days for each casting (produced in parallel). The cost of casting each of the A-posts from the master pattern was SEK56000 (£3752).

6.5.9   Phase 9 – Finishing work to A-posts and assembly to the car

The final castings were sent from Tritech to Volvo to be inspected, finished, and assembled to the concept safety car. The finishing work took 1½ weeks and involved heat treatment for correction, milling and drilling for holes and planes, and some minor finish adjustments. This finishing work resulted in the good quality parts required. Estimated cost for finishing work was SEK80000 (£5360) for each A-post. Figure 6.8 shows the final functional car with the A-posts in place.

Fig. 6.8 Functional car with A-posts

6.5.10 Phase 10 – analysis of results

The early phases of the project (phases 1–6) showed that investment casting of stainless steel into shells created from SL QuickCast™ patterns was the best available route for production of the A-posts. The final A-posts were produced in the times and costs given below, which compared very favourably with the alternative of machining. Estimated costs and times for machining indicate that milling would not incur more time or cost, however, it would very difficult due to the complexity of final geometry and stresses created in the material during milling. Table 6.2 summarizes the times and costs required to manufacture the A-posts.

Table 6.2 Summary of times and costs required to manufacture the A-posts

Process	Time (weeks)	Cost for first A-post SEK (£)	Cost for second A-post SEK (£)
SLA master	1	39 000 (2613)	42 000 (2814)
Shell and casting	4	56 000 (3752)	56 000 (3752)
Correction	½	20 000 (1340)	20 000 (1340)
Milling planes and holes	1	60 000 (4020)	60 000 (4020)
Total	**6½**	**135 000 (£11 725)**	**138 000 (£11 926)**

Conclusions and recommendations

The complete functional SCC was designed and successfully manufactured at the Volvo Concept Centre in Gothenburg. Before this project was undertaken, Volvo Cars were presented with the problem of finding a suitable manufacturing route for the new, novel A-post design. While traditional routes such as milling were known, the possibility of using QuickCast™ patterns as sacrificial patterns for investment casting had not been fully understood and considered. The suggestion by RMC for this alternative route allowed Volvo to experiment with a different manufacturing approach and make contact with an expert in the process, within the PAG.

The initial test casting indicated that the route of using QuickCast™ patterns for investment casting was suitable for producing one-off castings and that stainless steel should be used. The process chain for producing the final A-posts showed that the lead time from CAD file to finished part was 6½ weeks, which was comparable with the traditional alternative of machining. Additionally, the costs involved were comparable with machining, however, the casting method ensured that the final parts included less stress induced by the manufacturing process. While stereolithography patterns were used in this case, it is possible that selective laser sintering of polystyrene using either EOS or DTM/3D Systems machines could be used to produce the sacrificial pattern.

Staff at Volvo cars are now more familiar with the investment casting process, using QuickCast™ master patterns, and may find other suitable applications where other less suitable approaches may have otherwise been taken.

The functional safety concept car was unveiled in Seville in November 2001 and has been test driven by various members of the press and media. The car has won a number of international automotive design prizes including Autocar magazines Design award in the Autocar Awards

2001 which quoted the car's 'revolutionary see-through A-pillars for superior vision, and state-of-the-art technology for improved driving and pedestrian safety'.

References

(1) **Jacobs, P.** and **Kennerknecht, S.** Stereolithography 1993: Epoxy Resins, Improved Accuracy and Investment Casting, SME Rapid Prototyping and Manufacturing Conference 1993, May 11–13, Dearborn, Michigan, USA.
(2) **Hague, R. J. M.** The Use of Stereolithography Models as Thermally Expendable Patterns in the Investment Casting Process, Ph.D. Thesis submitted to the University of Nottingham, January 1997.
(3) **Hague, R.** and **Dickens, P.** Finite Element Analysis Calculated Stresses in Investment Casting Shells Using Stereolithography Models as Patterns. Proceedings from the 5th European Conference on Rapid Prototyping and Manufacturing, June 4–6, 1996, Helsinki, Finland.
(4) **Jacobs, P. F.** Stereolithography and other RP and M Technologies, ASME Press, 1996, ISBN 0 87263 467 1.
(5) **Stierlen, P., Dusel, K. H.,** and **Eyerer, P.** Investment Casting: PR Patterns for Metal Prototypes, Proceedings from the Time Compression Technologies Conference, 13–14 October, 1998, Nottingham, UK.

N Hopkinson
Rapid Manufacturing Consortium, formerly De Montfort University, now Loughborough University, UK
J Almgren
Rapid Prototyping Centre, Volvo Cars, Göteborg, Sweden

7

Rapid Prototyping with Vacuum Investment Casting

R Minev

Abstract

The rapid vacuum investment casting process is recently successfully used to produce metal parts from non-Fe alloys (Al, Zn, Cu, Mg, etc.) using different rapid prototyping (RP) patterns (SLS, SLA, FDM, Sanders Process, etc.). The process could be classified as precision expandable pattern, block mould, investment casting. RVIC is capable to produce both thin wall (minimum 1 mm section on 10^4 mm^2) castings and thick wall castings (up to 15 kg Al-alloy) if the suitable design and technological parameters are chosen considering the appropriate solidification strategy for the casting process.

Introduction

The new approach to casting manufacturing engineering is relevant with the development of a vacuum investment casting process. During recent years, this process has rapidly expanded in respect of new practical applications, accelerating of development time of new products from CAD design to the real product, as well as the utilization of new alloys. The close interaction and co-operation of vacuum casting with rapid prototyping (RP) processes such as SLS, SLA, FDM, etc. (1), further promotes the ability and competitiveness of the process.

Although some new methods and materials for rapid production of metal parts, such as LENS-technology, SLS-LaserForm, etc., were recently developed (2), the final product to come out suffers from all well-known disadvantages of powder-metallurgy. Rapid investment casting (RIC) is the only available and reliable process that gives predictable structure and properties to the final product. The fast development of vacuum RIC, in terms of improving the reliability of the process, the usage of different materials for producing patterns with high accuracy, and the ease of the process, has left behind lots of unsolved metallurgical and technological problems.

7.1   The problems of RIC

Previously the problems of casting design (including the design of the feeding system and solidification history), as well as defining the casting parameters for different alloys, were the focus of research work. The fast blooming of RP methods has concentrated attention on the problems of developing new model materials and burn-out procedures for investment casting. The casting process itself is considered less important with problems easy to solve. The result is that, although we have a large variety of RP methods and materials for casting modelling, the foundries are very restricted in their implementation. The main reason for this is that there is not much knowledge on the limitations of their usage. In other words, there is a huge gap between RP development and the state-of-the-art in investment casting capabilities to produce accurate and complex castings for a reasonable price (3). Some of the unsolved problems could be summarized as follows.

7.1.1   The solidification strategies
For the conventional casting methods, the problem of shrinkage is amongst the main technological problems. A significant amount of experience exists on how to chose the 'directional' or 'simultaneous' solidification strategies (4–5). The wall thickness and complexity of the casting are the main criteria for this choice, but also the type of casting alloy has to be considered. In modern investment casting processes, the application of a vacuum and an overpressure has been introduced to reduce the negative effect of the shrinkage on the quality of the castings. However, these measures are not capable of completely solving the quality problems (sink marks, structural defects, etc.). Again, both strategies of 'directional' or 'simultaneous' solidification could be applied to reduce the shrinkage and porosity. Unfortunately, the criteria for their usage are not yet clearly determined. Most of the RP products to be cast in metal are thin walled parts, with small or medium volume. Usually they can be produced with the strategy of 'simultaneous' solidification. At the same time, there is a huge lack of experience in the 'directional' solidification approach applied to vacuum investment casting. For this reason, the application of RP is limited to small and thin wall castings with significantly complex shapes. The alloys in use in RP are also limited for the same reason. Mainly non-ferrous alloys with a small range of solidification temperatures are cast with the vacuum investment casting method. Consequently, we have an understanding among the users, that RIC processes are not capable of producing large- and thick-walled elements.

7.1.2   Dimensional accuracy
In RP casting technology, we can observe two or three thermal processes, which contribute to possible errors in dimensions and shape. These are:

- the process of sintering or curing the polymer patterns (SLS, FDM, SLA, etc.);
- the process of building or the infiltration of the patterns with wax (SLS-CastForm, Silicon, or Epoxy Tooling); and
- the metal casting process.

RP methods for producing the plastic patterns are relatively well explored in terms of the accuracy. The only exemptions are the most popular materials in the investment casting industry – wax infiltrated polystyrene (CastForm) and fused deposited wax (ThermoJet and Sanders Process). Some very important technological aspects of these techniques (support structure and wax-infiltration) are still under development. Also, there is not much information on the shape stability and dimensional accuracy of the vacuum investment casting process itself. For example, the wax infiltration of the polystyrene mould patterns widely affects the accuracy and quality of the final patterns. There are some guidelines on the temperatures and times to be used during the process but the shape, size, and complexity of the parts must be taken into account.

7.1.3   The usage of overpressure in vacuum investment casting

The overpressure is an essential element in the strategy of filling the mould cavity with molten metal. It enables the metal to quickly fill the most intricate sections of the cavity and ensures a significant compensation against the metal shrinkage. It is very useful when the 'simultaneous' solidification strategy is applied reducing the porosity and giving a better resolution to the casting. When the 'directional' solidification approach is applied the casting will need better venting. Overpressure could even interfere with the 'take out' of gasses.

7.1.4   Properties of the casting

Without the knowledge of the essential mechanical and some specific physical properties of the parts produced using RIC, the transformation from RP to rapid manufacturing is unfeasible.

7.2   The rapid investment casting technology

The case studies that are covered in this Chapter are dedicated to the application of the vacuum rapid investment casting (VRIC) technology to produce thick-wall metal RP components of significant weight and size. The VRIC technique (Fig. 7.1) was designed and mainly used for small thin wall components with complicated shapes **(6)**. The process (of which technical data are presented in Table 7.1) could be classified as precision expandable pattern, block mould, investment casting.

Overpressure up to 1400 mBar

Vacuum -1000mBar

Fig. 7.1 Vacuum investment casting machine MPA 300 (MCP equipment)

Table 7.1 Some technical specifications of the process

Alloys	*Al, Zn, Cu, other non ferrous alloys*
Accuracy	Up to 0.1% on linear dimensions
Pattern materials	SLS-CastForm, FDM-ABS, ThermoJet-Wax, Sanders – pattern master Wax, etc.
Maximum flask size	Diameter 350 × 500
Minimum flask size	Diameter 125 × 250
Maximum melting temperature of the alloy	1300 °C
Crucible capacity	3 litres
Time from STL file to a real part	Minimum 48 hours
Induction heater power	25 kWt
Vacuum	−1000 mbar
Investment powder	MO28 or relevant
Overpressure	Maximum 1500 mbar

The essential steps in the manufacture of the castings may be summarized as follows:

- production of expandable patterns by some of the existing RP processes (SLS, SLA, FDM, etc.);
- investment of the patterns to form a one-piece refractory mould **(7–8)**;
- pattern elimination and high temperature firing;

- casting; and
- finishing.

In the process of casting the applied vacuum of about 1000 mbar draw out the air from the mould cavity through the embedding ceramic. At the same time the bottom stopper in the crucible is opened and the molten metal fills the mould cavity, an overpressure of protective (nitrogen or argon) gas is applied to the crucible. The vacuum and overpressure facilitate the metal flow, degassing, and shrinkage compensation of the casting during the solidification.

Figure 7.2 represents some typical thin wall castings made with this process of LM6 aluminum alloy using SLS CastForm patterns. The wall thickness of the casting is in the range of 2.5–1.5 mm and the surface area of the biggest shield-like castings is $0.16*10^6$ mm^2. The time for making a set of 15 castings was three days. The dimensional accuracy was within the range of 0.2 per cent.

Fig. 7.2 Thin wall aluminum alloy castings produced by RIC

However, as we will see in the examples quoted below, thick-walled, massive castings introduce specific problems that are difficult to solve because of a lack of knowledge and experience.

7.2.1   Case study 1 – the pedal frame

The first case study is about a part with 'L-frame' shape that holds three hydraulic cylinders and a lever system to move the pistons in them (Fig. 7.3). The unit was designed for a sports automotive application. It went through a number of modifications during the prototyping stage, when 30 pieces of the unit were completed over a one-month period.

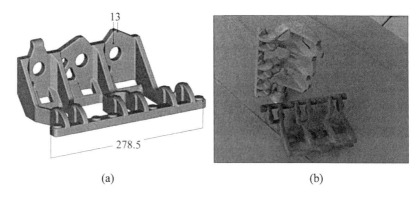

(a)   (b)

**Fig. 7.3 The STL model (a) and a CastForm (wax-infiltrated polystyrene)
pattern and metal parts with the gating system (b) of the pedal frame
(courtesy of Vale Castings Limited)**

The mass production of the unit was foreseen to be accomplished by the gravity die-casting technique. A LM25 aluminum alloy was chosen for its approximately one per cent Mg content, which enabled an effective precipitation hardening heat-treatment to be conducted. The cast surfaces were subjected to hard anodizing to improve the aesthetic appearance and environmental stability of the unit.

Machining of the functional surfaces to join the cylinders and levers was done by means of five axes CNC machining. The accuracy achieved for most of the dimensions of the castings were within the range 0.3–0.5 per cent. The machining of the surfaces, being offset from one to another, required a multi-datum (using several reference surfaces) approach to distribute the errors and achieve the most accurate assembly dimensions. The machining allowance varies within the range of 0.3–1 mm for the different surfaces.

7.3   The casting aspects of the component

The analysis of the component shape leads to the decision to locate the 'L-shaped' frame shoulder vertically and cylinder placement shoulder horizontally [Fig. 7.4(a)]. The vertical shoulder could be used as a 'natural' branched out feeding system for the casting. The above-described position has also been chosen for the die-casting of the component. Thus, the investment casting of the prototypes would simulate the mass production technology of the component. However, for the vacuum investment casting this first approach proved to be, by far, not the best. Some of the important disadvantages of this position were as follows.

- Necessity to use bigger flasks (diameter 350 mm instead of 250 mm). This increases the volume of the embedding mass necessary to produce one part by more than 70 per cent.
- Usage of the bigger flask reduces the efficiency of the vacuum, because the ceramic walls were thicker, as well as the reliability of the process by increasing the risk of cracking and failures.
- To increase the efficiency of the process we arranged two patterns per casting tree. This significantly increased the amount of metal to be cast and changed the thermal conditions

of cooling the mould and metal solidification. The overheating of some spots (especially the T-shape 'thermal knot' on the frame) caused rough dendrite texture on the surface and even short cracks (see Fig. 7.5). The overheating of the mould also reduces the permeability of the ceramic and reduces the degassing of the casting surface. The casting parameters for the pedal frame are shown in Table 7.2.

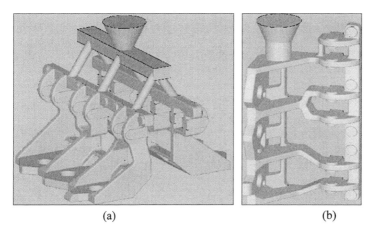

(a)   (b)

**Fig. 7.4 The casting design for the first (a)
and second (b) approach of the pedal frame castings**

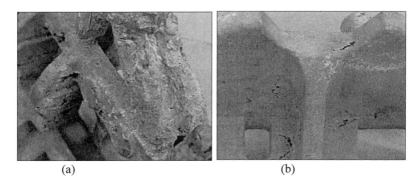

(a)   (b)

**Fig. 7.5 Casting defects on the pedal frame: trapped air
and porosity (a) short cracks on the local hot spots (b)**

Table 7.2 The casting parameters for different parts

	Pedal frame, 1st approach	Pedal frame, 2nd approach	Piston	Welsh Quality Award
Tmetal	Tm + 60 °C	Tm + 30 °C	Tm + 60 °C	Tm + 30 °C
Tflask	50 % Tm	40 % Tm	170 °C	170 °C
Vacuum	−1000 mbar	−1000 mbar	0 mbar	−1000–0 mbar
Overpressure	+1400 mbar	0 mbar	0 mbar	0 mbar
Overpressure delay	4 sec	–	–	–

The causes for the observed defects (9) were the relatively high temperature of the flask and molten metal as well as the large quantity of the metal per flask. They facilitate a low speed of freezing, overheating of the surface with a consequent dendritic structure of the metal, and short cracks in the 'thermal knots'. The application of the overpressure averts the degassing and causes the gas porosity observed in the thicker sections of the castings. In fact, the cooling history of the casting corresponds to a directed solidification of the metal, but the design and technology parameters did not facilitate the controlled freezing and assisted degassing of the metal. Two ways of solving the problems were possible: change the design and casting parameters toward the directional solidification approach; or providing conditions for the metal to solidify faster and simultaneously in the die cavity. The first way proved to be unsuitable because it required an additional increase of the volume of the casting and a consequent increase of the flask size, embedding mass, and time to burn out the patterns. The second method was chosen, although the thickness of the sections of the pedal frame was relatively high and there was no existing experience, and so no guarantee of success of the 'simultaneous freezing' of the casting. The problem of defining the criteria of the casting configuration, to prove its suitability for the simultaneous solidification in vacuum investment casting process, was still not solved. As it will be shown later, a casting with an appropriate configuration was used to determine more accurately the section thickness of the castings capable of being 'simultaneously frozen'. Nevertheless, the usage of simulation software is imperative and could completely solve the problem in the future.

The second approach, to produce pedal frame prototypes, was aimed to bring about appropriate conditions for the metal to solidify quickly throughout the entire volume of the casting and to enable easy degassing. The position of the part in the flask, was with the longest dimension paralleled to the flask axes, with the gate and pouring basin attached to the side [Fig. 7.4(b)]. Only one part per flask was cast, thus reducing the volume of the freezing metal, the embedding mass, and time for burn out of the pattern. Reducing the temperature of the flask and temperature of the molten metal reduced the overall time of solidification. No overpressure was applied, since it was proven to interfere with the degassing of the metal from the feeding system. The volume of the feeding system was reduced to seven per cent of the total volume of the casting. The technological parameters of the casting are shown in Table 7.1.

The implementation of the casting design and parameters in the second approach gave excellent results. The surface quality of the casting was very good and no porosity and sink marks were observed. The dimensional accuracy of the casting was also within the limitation of the process, e.g. 0.4 per cent (maximum relative error).

7.3.1   Case study 2 – the two stroke engine piston

In the second case study, we will show a successful implementation of the guided solidification strategy to produce a thick section casting by the VRIC process. The part (Fig. 7.6) was a piston with a flange, and sophisticated ring cavity connected with long channels for cooling liquid and for a thermocouple. The internal cavity of the piston was also very deep and was foreseen to be EDM shaped at the bottom. The piston was design for an experimental two-stroke aviation engine. The LM25 aluminum alloy was also picked for this casting.

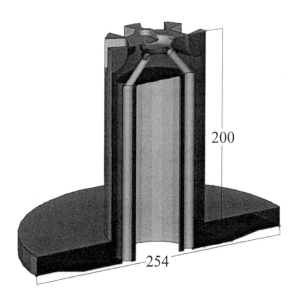

**Fig. 7.6 Section of the piston, showing the internal cooling cavities and holes
(courtesy of John Clark Engineering)**

The complexity of the cooling cavity required the use of soluble casting cores. Another solution was to separate the STL file of the casting in two parts – the body and a small cylindrical cap, which could be cast separately and then joined together (Fig. 7.7). This option seemed more reliable and easier to fulfill. The diameters of the hole in the body and the cap were calculated to fit after heating the body (to the solution heat treatment temperature 800 K) and cooling the cap (to the temperature of about 200 K). Then the two parts were joined together. During the equalizing of the two temperatures the stresses on the interface surface exceeded the yielding point of the aluminum, hence some plastic deformation and forge welding between the surfaces took place leading to a completely sealed assembly.

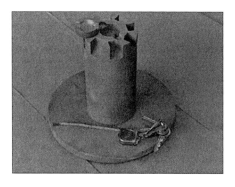

**Fig. 7.7 The piston was cast in two parts, the body and a cap,
which was later press-joined together**

The volume and wall thickness of the casting involves a 'directional strategy of solidification'. Two approaches to design the casting were attempted. The first one was more traditional. The part was placed horizontally in the flask and some risers were added to the piston and flange [Fig. 7.8(a)]. The advantage of this design was that the big internal cavity is placed horizontally in the flask and would be easily filled with ceramic slurry during the embedding process. However, this position of the casting required the usage of a bigger flask and also the volume of the effective additional risers was rather high. The total weight of the casting reached 11.5 kg, but the porosity and shrinkage was still not effectively brought out from the body of the part to the sacrificial risers. Any further increase of the total volume of the casting would dramatically exceed the capacity of the casting equipment, which is a maximum of 7 kg of Al-alloy. To overcome this problem part of the metal was melted outside the vacuum casting machine and poured into the cavity before the metal was completely solidified.

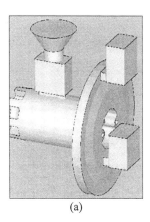   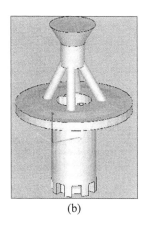

(a)   (b)

**Fig. 7.8 Different casting design for the engine piston:
(a) horizontal placement; (b) vertical placement**

In the second approach to designing the casting, the part was positioned vertically in the flask [Fig. 7.8(b)]. The feeder head, sprues, and risers were attached symmetrically to the part. The total volume of the casting decreased by 20 per cent, and the effectiveness of the degassing was very good. The only disadvantage of this approach was that the internal cavity of the piston was open to the bottom of the flask and embedding mass did not fill it easily, leaving trapped air. A ceramic core was used to preclude the casting defects that appeared in the bottom of the cavity. It was produced from the same ceramic material used for embedding the flask. The pattern cavity was filled with slurry prior to the final embedding. An extension core print with a draft angle was formed to fix it into the ceramic. After solidification of the core, all additional elements of the pouring system were added and the pattern was turned up side down and embedded into the flask. Then the ceramic core was joined together with the main ceramic to form the mould. The operation described above did not increase the labor and time needed to produce the mould, but dramatically improved the quality of the internal cavity of the casting. The casting parameters are shown in Table 7.2. The application of overpressure was not necessary and the degassing of the casting was better without it. As far as the vacuum is concern, the analysis of the kinetic of casting solidification shows that in the case of directed solidification of massive, thick wall castings it could have a negative effect. Indeed the piston castings produced using of a vacuum of -1000 mbar showed significant surface porosity on the cylindrical surface. The reason for this phenomenon is simple. The application of the vacuum alters the solidification conditions nearer to the case of fast solidification. When the wall thickness is small, the sucking force through the mould walls is a positive factor facilitating the closer contact of the metal to the cooler mould surface. Thus, helping the process of solidification to occur in a non-equilibrium condition with a formation of multiple grain nuclei and a fine grain structure of the casting. On the other hand, the vacuum force facilitates the cavity filling and degassing of the metal. When the wall thickness is considerable (as in the described cases) the vacuum can play a negative role. The directional solidification of the casting from the surface to the core requires good feeding of the interface layers with molten metal from the core, which will fill the gaps between the already frozen grains. If a vacuum was applied at the same time, it would 'suck' the cavities between the solidified grains to the surface leaving them unfilled with metal. This 'sucking' effect must not be literally comprehended as a movement of the cavities. It has to be perceived as rather fast cooling of the surface layers, which does not allow the molten metal to move and fill the pores. In other words, if because of its configuration the casting requires directional solidification it is not appropriate to use any methods that cause an increase in the speed of metal freezing.

Another lesson from this case is always to consider the directed solidification, not only from the global point of view of the entire volume, but also from the point of view of local solidification and the consecutive freezing of the layers, starting from the metal/mould interface to the center of the section. Of course, the elimination of the vacuum in the mould cavity will deteriorate the degassing of the casting and will require additional measures to support it. In the described case, it was a set of additional risers, attached to the flange and opened to the flask surface. The described dilemma of whether and when to use the vacuum in the ceramic mould is playing an essential role in defining the strategy of the investment casting process. It also sheds light on the difference of principle between the conventional and gravity investment casting on one hand, and the vacuum investment casting on the other. In the conventional casting processes, the permeability of the mould ceramic is one of its major properties. It plays an essential role, facilitating the process of degassing the molten metal. In the vacuum investment casting the porosity and air conductivity of the ceramic is much less and must be compensated for, by the vacuum, applied from the outside of the mould. The vacuum increases the airflow through the embedding mass. The decision of whether or not to apply a vacuum must be taken very carefully considering the wall thickness and the configuration of the casting so that the successful filling of the cavity can be achieved.

7.3.2   Case study 3 – the Welsh Quality Award

The definition of appropriate criteria for choosing the directed solidification and the usage of the vacuum is very difficult in the RP business environment because the time and facilities to set up an experiment are very limited. In this case study we were fortunate to work on a casting, which could be used as a test sample to define such criteria. That was a casting with a gradually decreasing cross section from 5 mm^2 to $5*10^3$ mm^2, similar to the standard test samples widely used for determination of the castability of the alloys. In addition, the casting (Award Prize, Fig. 7.9) had enhanced requirements to the surface finish and a lack of porosity. The part was cast in different conditions: with, and without vacuum and overpressure, and using different flask temperatures. Another interesting innovation was the use of the vacuum only during the pouring of the metal (for a time of three seconds) to facilitate the filling of the cavity. Then the vacuum chamber was aerated and metal solidified in directional solidification conditions.

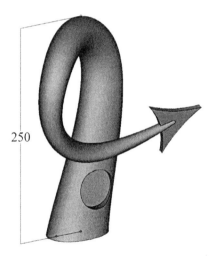

250

Fig. 7.9 Welsh Quality Award prize

The observation of the surface porosity of the casting gives an impression of the critical wall thickness (or the area of the section) under which a successful 'simultaneously' cast metal part could be produced. The temperature of the flask and the temperature of the metal affected the critical dimension (the section where porosity appears on the surface). It varies from $2.5*10^3$ mm^2 to $1.5*10^3$ mm^2 when the temperature of the flask was changed from 200 °C to 300 °C and reduces when the pouring temperature of the metal T_m increases. A summary of the conditions to form a successful casting free of defects using the vacuum investment casting machine is given in Fig. 7.10.

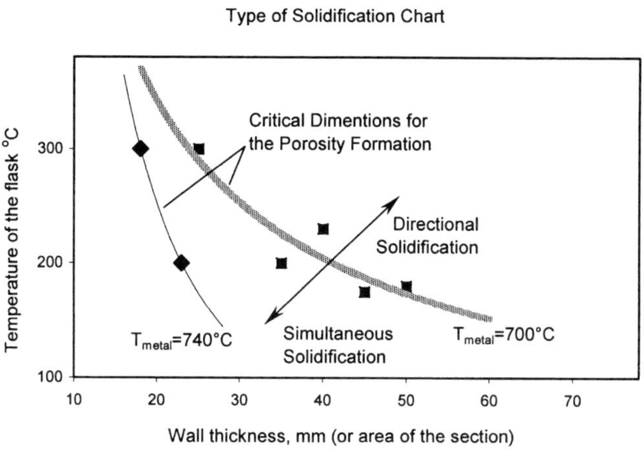

Type of Solidification Chart

Fig. 7.10 A chart for defining the appropriate casting strategy

In this figure the results from the previous case studies (pedal frame and piston) were also added. The most important parameters of the casting process that affect the way the metal solidify: T_{flask} (temperature of the flask) and the free of pores wall thickness are presented on the Y- and X-axis of the co-ordinate system. This diagram could be successfully used to resolve the most suitable casting strategy ('directed' or 'simultaneous' solidification) as well as the appropriate flask temperature to make a good quality casting. The zone above the curve represents the conditions when a directional solidification approach will give a non-porous casting while the area below the curve the condition for the simultaneous fast freezing approach to be used. Most of the present experience falls below the curve, while the current presentation covers above the curve and aims to prove the capability of the process and to extend the limitations of the process. The influence of the pouring temperature of the metal (T_m) to the critical parameters was not deeply researched but the diagram shows some of the existing data. There is a well-established tendency of shifting the boundary conditions to the smaller dimensions when T_m increases.

Conclusions

RIC is capable of producing both thin wall (minimum 1 mm section on 10^4 mm^2) castings and thick wall castings (up to 15 kg Al-alloy), if the suitable design and technological parameters are chosen considering the appropriate solidification strategy for the casting process.

The maximum size of the castings would be approximately 300 × 300 × 600 mm. The minimum size of the castings according to our experience is 5 × 5 × 8 mm. The process has the possibility to cast cylindrical holes with diameter to length ratio of 0.1 and multiple ribs with depth to thickness ratio of 0.2.

The possible time to produce a part starting from the CAD file to the metal casting is 48 hours and depends on the size and complexity of the shape.

The accuracy and surface finish of the parts depend mostly on the RP pattern quality. They are within the range of 0.1–0.2 per cent tolerance on the linear dimensions and N7–N10 (ISO 1302) surface finish.

References

(1)   **Jacobs, P. F.** Stereolitography and other RP&M Technologies, from RP to Rapid Tooling, SME, Dearborn, Michigan, 1996
(2)   **Pham, D. T.** and **Dimov, S. S.** Rapid Manufacturing, Springer, London, 2001
(3)   Incorporation of New Technologies in the European Precision Foundry Industry, SARE – Thematic Network founded bu EC under 5th Framework Program, www.sare-network.org, 2000
(4)   **Beeley, P. R.** Foundry Technology, London Butterworths, 1980
(5)   **Gradinarov, A.** Foundary Technology, Rousse University, Bulgaria, 1981
(6)   **Fretwell, R. A.** MCP Metal Casting System, Manufacturing Engineering – Newsletter, MEC – Cardiff University, Autumn 1999
(7)   Manufacturing Engineering Center, Part 2, RP at the MEC – in-house Technologies, Engineering, 2001
(8)   Hoben International, Jewelry, www.hoben.co.uk
(9)   **Schey, J. A.** Introduction to Manufacturing Processes, McGraw-Hill, 1987

R Minev
Systems Engineering Department, Cardiff University, Wales, UK

8

Direct Shell Sand Rapid Prototyping

C Ryall, G J Gibbons, and **R Hansell**

Introduction

Casting is one of the most important production methods and plays a role in 90 per cent of all finished manufactured products. The speed at which castings can be generated has improved with the advent of automatic moulding processes. One basic requirement of the process has always been a pattern. Generation of the pattern in some instances can take many months and does not guarantee a successful product or sound casting. The advent of rapid prototyping and high-speed machining has partially addressed the problem of the time required to produce a pattern. The use of these techniques has seen a steady reduction in the lead times associated with the generation of both prototypes and production equipment. However, if an error in design or poor performance of the casting is discovered, costly modification or re-manufacture of the pattern equipment may be necessary and the associated costs can be very high. Manufacturers are, therefore, reluctant to commit to tooling if there is any doubt with respect to the performance/design of a casting.

The development of RP systems has seen the evolution of machines capable of manufacturing moulds in both ceramic and resin bonded sands. These machines are ideally suited, where the design of a casting is still fluid, to produce prototypes for evaluation purposes. Manufacturing times for the production of the most complicated castings can be reduced to weeks or even days without the need for a pattern. The process can also make the manufacture of single items economically viable.

This Chapter will describe the laser sand sintering process, its limitations, the problems associated with its use, and the need to educate potential users to its requirements through industrial case studies.

8.1   Rapid prototyping

Rapid prototyping is a relatively new term for the generation of three-dimensional models without the need for machining or tooling. The advantage of the process over the more conventional techniques, such as machining or hand skills, is the speed with which very complicated geometry can be reconstructed.

8.1.1   Data requirements
To use rapid prototyping, a three-dimensional CAD model is required of the desired form. This then has to be translated within a CAD system into the STL file format. The STL file is an electronic image of the desired form. Once generated, the STL file is electronically sliced into layers of predetermined thickness, usually between 0.1–0.25 mm.

The rapid prototyping machines use a variety of different techniques to reconstruct this data, layer by layer, into a tangible physical model.

If problems occur it is usually because of poor quality CAD data or because the requirements of the STL file are unknown to the user. The data has to be perfect, unlike that required for machining where small discrepancies can have little or no effect on the final form.

8.1.2   Laser sintering
Laser sintering (LS) is one of many commercially available RP systems. The process was developed by the University of Texas and was subsequently commercialized by the DTM Corporation. Whereas stereolithography (SLA) uses an ultraviolet curable liquid resin, the SLS process uses powdered materials. This is one of the major advantages of the system, because, in principle, a model could be built in any fusible powdered material. Currently the range of materials includes nylon, glass filled nylon, sand, and metals.

The laser scans each slice fusing the powder, to the previous layer. As with all of the techniques, the model is built on a table, which lowers after completion of each slice. A fresh layer of powder of the required thickness is then spread across the top of the build area and the process is repeated. Excess powder remains in place around the model to act as a support during the build. On completion of the build the excess powder is brushed from the surface of the model.

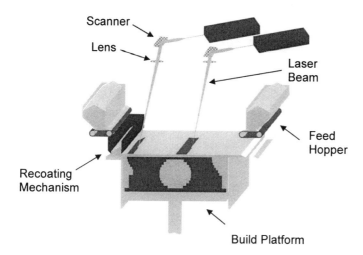

Fig. 8.1 Schematic of the EOS S700 laser sintering process

8.1.3   Laser sintering of sand – direct croning

The direct sintering of sand relies on the use of resin coated sands. The application of heat from a CO_2 laser locally heats the sand grains and bonds the material together. The idea of putting resin-bonded sands into sintering equipment was patented by Dr Wendt of AC Tech in 1994. After sintering the moulds and cores have all excess unsintered sand removed from the surface of the geometry. This can be time consuming, as complicated cores can be very fragile. Once the sand has been removed, the parts are flame treated with a small blowtorch. This creates a hard skin on the surface of the component giving it far higher resistance to surface damage. The cores and moulds are submerged in a bed of glass beads and heated to 180 °C. The beads give support to the parts, as the resin softens on reheating before final cure. There are currently two manufacturers of machines available for the sintering of sand; 3D Systems and EOS. 3D Systems sintering machines are capable of sintering a wide range of materials including polymers, resin coated sand, and resin coated metals. The build envelope of their range is 380 × 33 × 420 mm. EOS produces a range of dedicated material machines. Their S700 machine is dedicated to sand with a build envelope of 720 × 380 × 400 mm.

The RP&T centre uses an EOS machine for the production of sand cores and moulds.

8.2   Main users of sand sintering

Most general RP bureaux rely on either, direct investment casting of models from the various RP systems, or in providing models to pattern makers for conversion to patterns. Both routes can be fairly rapid, but this is dependent on the model geometry. A number of UK pattern makers do not believe these routes have any cost and time benefits over high-speed machining of pattern equipment (1). The machining of patterns also has the advantage of being both robust and capable of high numbers of castings without the risk of excessive wear.

The application of sand sintering is not currently mainstream within the RP industry, there are relatively few bureaux offering this as a service. This is probably due to the requirement that the user has to have knowledge of pattern making and methodizing, or to work closely with someone with this knowledge.

The main users of the systems to date have been automotive companies and their suppliers. BMW, VW, Porsche, Ford, and Mercedes have their own in-house facilities, which use both internal and external expertise. The advantages to these companies are obvious when considering the cost and time implications in designing and producing new cast automotive components. BMW, Ford, and VW have both produced prototype cylinder heads in a fraction of the time and cost of traditional routes (2–4). There are a range of pattern makers and casting companies who have decided that this technology has distinct advantages for them. One of the most startling examples of a pattern makers growth after purchasing this type of equipment is Erreci Rapid Casting in Italy (4). The company is part of a group of companies, which has at its disposal high-speed machining centres and traditional pattern making skills.

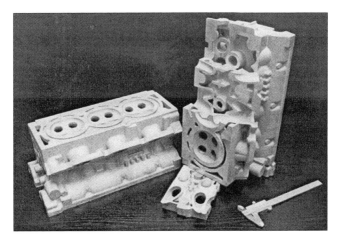

Fig. 8.2 A cylinder head produced by laser sintering.
Courtesy CAD-CAM-Team-Becker

After purchasing an EOS S700 in July 1997, the machine was soon overwhelmed by its own success. Within 15 months Erreci Rapid Casting had purchased four additional S700 systems.

The range of castings produced by these companies is very broad covering, multi-cylinder and single cylinder heads, inlet and exhaust manifolds, engine blocks, hydraulic valves, and many other complicated forms.

8.3   The advantages and disadvantages of sand sintering

8.3.1   New product design

The systems main advantage is its potential to provide castings in as little as a day from provision of suitable data. As a wide range of materials can be cast into these moulds the casting can be considered a 'technical prototype' without the need for tooling. This short lead-time allows greater design flexibility. The castings can be tested for fit, flow, stressed, used for machining trials, and as an aid for the production of jigs and fixtures. The performance of the casting and final product can then be fed back into the design of the component long before there is any requirement for tooling.

In some instances, a series of prototypes can be made for the cost of a prototype tool. This allows the design to be refined fully before committing to production tooling.

Other benefits include the refinement of feeding and gating systems to ensure sound castings. The process can also be used to generate complex geometry's where previously the number of castings was economically unfeasible. The main problem to date has been convincing companies that this extra cost incurred early in the design cycle has value.

In instances where the design is relatively simple there is not much to be gained. As the complexity of the geometry increases the case for the use of this process becomes stronger.

8.3.2   Mould data

As mentioned previously, the first major drawback of the equipment is the need for an STL file of the mould geometry. Most files provided by engineers are of the finally machined component and are not castable. Modification of the CAD data is, therefore, essential. The RP bureaux that offer this service are either pattern makers or foundries and are in a position to offer this as a service. Alternatively, the more mainstream bureaux offering this service have brought in the necessary expertise.

It should be noted that the process does not require draft or a clear line of draw for pattern equipment and can produce otherwise previously unmanufacturable moulds. This has no benefit where the product is expected to go into series production later in the design cycle.

8.3.3   Surface finish

One problem with the process is poor surface finish. As the laser scans the slice some grains of sand remain unsintered while others become part of the mould or core. With a traditional mould all grains of sand are aligned to the surface of the pattern or core box. Quoted figures from EOS state an RA value of between 50–80 μm for an aluminium casting produced in one of its moulds compared to 50–70 μm for a casting produced in a normal green sand mould **(1)**. Where an improvement in the surface finish of the casting is required, for example exhaust and inlet ports, mould coatings have to be used.

8.3.4   Resin, out-gassing

The current sand uses a phenolic binder, due to the relatively low thermal input of the laser, the resin content of the sand is fairly high being in the order of five per cent. This however, can cause problems with gas in the mould during casting. Tests at The RP&T Centre have shown the sintered sand to have similar permeability to the Betaset system using AFS 60 sand. For most geometry this is not considered to be too much of an issue, but where a core is nearly completely enclosed by metal, as with an exhaust manifold or hydraulic valve, additional venting is advisable.

Venting can be added one of two ways, in the first case, vents are added to the CAD model or STL file, which are subsequently built into the core or mould by the machine. 'Shelling' of an STL file is easily achieved as software automatically hollows the geometry in a matter of seconds. This also has the advantage of reducing the build time for the geometry as less sand has to be sintered. More commonly, moulds and cores are drilled before casting, especially where removal of the unsintered sand is not feasible due to the fragile nature of the semi-cured sand and geometry. A hollow core for a manifold is shown in Fig. 8.3.

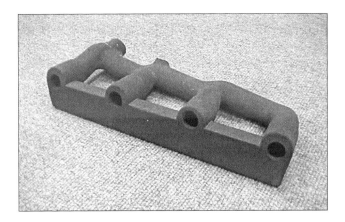

**Fig. 8.3 Hollow core stick produced on the EOS S700 machine.
Courtesy NPL Technologies Limited**

8.3.5   Accuracy

Accuracy of the process is determined by the following factors; the scanning system, the grain size of the sand, layer thickness, and post build operations, which can all have an effect on the final result. The scanning systems used on most RP machines are accurate to within approximately ±0.1 mm. The sand's maximum diameter should be less than 200 μm as all new and used sand is sieved before use. Accuracy is, therefore, in the order of 0.3 mm in any given direction. The main source of inaccuracy can occur during post build operations, as the sand is being surface treated with a micro-blow torch it is possible to raise the temperature of a section to the point where the resin starts to soften. Distortion of the geometry may then occur and in some instances this is very difficult to detect. This, therefore, relies on the knowledge, skill, and experience of the operator.

8.3.6   Break even point

It is generally considered that the break-even point for sand sintered moulds compared to traditional machined pattern equipment is in the order of 1–10 castings; this though depends on the geometry. Thereafter, a hybrid route is used where complicated sections of a mould are sintered and traditional pattern equipment is used for the simpler or much larger sections. Both these routes have been used on numerous occasions within The RP&T Centre. For these instances figures vary, due to the complexity and size of the geometry. Figures currently available vary from six to as much as one hundred castings. Therefore, traditional routes are more economical.

8.4   Case study 1

In June of 2001 The RP&T Centre was approached by a major industrial partner to prototype four sets of cylinder heads for a new V6 engine. Each engine requiring one left- and right-hand cylinder head for verification of the final design. The left- and right-hand heads were not symmetrical, requiring two different mould packs for casting. The data was provided for both of the cylinder heads in its positive form as opposed to that of the final cavity (mould geometry) into which metal would be poured and it was then necessary to model the mould sections using this data.

Staff from the partner company were seconded to The RP&T Centre to help with this process. Over a relatively short period of time, approximately five working days, the necessary mould sections were generated from the STL files using Materializes Magics software. WCM Patterns, a local pattern maker, offered guidance in mould design and was responsible for production of the castings. The final mould design for both the left- and right-hand cylinder heads consisted of 23 separate sand sections. Mould parts, such as the water jackets, were sintered before the final design of the mould had been produced cutting the final time scale. Additional time was saved by sintering multiple mould sections together to improve machine utilization. Details of each build can be found in Table 8.1.

Table 8.1 Sintering and post processing time scales
for the different mould components

Mould sections /cylinder head	Number of parts/build	Build (hrs)	Post processing (hrs)
Water jacket core (1 off)	4 off	4.5	3
Inlet cores (8 off)	4 sets	2.75	1
Exhaust core (8 off)	4 sets	2.5	1
Centre sections (4 off)	1 set	9.5	2
Cope (top – 1 off)	1 off	12.7	5
Drag (base – 1 off)	1 off	16	2

The first two moulds were cast within two weeks of the data being received by The RP&T Centre. Further pairs of castings were subsequently produced every three days thereafter until all eight had been manufactured. This was possible as shorter duration builds were run during the working day and longer builds were generally built at night. Each mould took approximately 41 hours to sinter and post-process. This time does not however take final cure of the mould components into consideration. This is normally run concurrently with the build operation after the first build is complete and, therefore, does not add to the duration of the project.

With this type of project there is considerable risk as there is no guarantee that casting metal into a mould will result in a sound casting. However, in this instance all eight cylinder heads were subsequently machined and used. The process has now been adopted in-house with the partner company having purchased its own equipment.

8.5   Case study 2

In May of 1998 Bevan Simpson Foundries approached The RP&T Centre to learn more about RP and how it could enhance their business. Later that year they introduced us to Anders Karlsson of VOAC Hydraulics Division, a subsidiary of Parker Hannifin Corporation specializing in mobile hydraulic valves. The group was looking for ways of reducing the lead times associated with the development of new products. The company was fully aware of rapid prototyping and its benefits. Their interest was purely in the potential of sand sintering, and over the course of this initial meeting it was agreed to build a prototype core to prove out the route. The geometry (Fig. 8.4) proved to be fairly challenging for us, due to our, then, relatively limited experience with the equipment.

Fig. 8.4 The core used for the initial trial with VOAC.
Courtesy VOAC Hydraulics Division

The core was subsequently cast and the valve measured for accuracy. The results were not as good as we had hoped, but were considered more than adequate for a prototype.

The main problem encountered was sagging of features during hot air gunning of the surfaces prior to final heat treatment. The use of this tool is no longer considered suitable and a micro-torch is used as it gives greater control.

VOAC decided to use the technique for the production of all prototype cores in the development of a new control valve, project Niagara. Over the next twelve months we produced ten or more different designs. With each set of new castings came revisions in the design or new cores, the latter occurring due to a change in law requiring greater safety functions.

During this period over one hundred and thirty six cores were produced, some of which are shown in Fig. 8.5. The average build time for these cores was less than one hour. Builds consisted of usually six or more cores at any one time. All of the cores were either cast in grey or nodular irons. The project was successfully concluded at the end of August 1999. The success of the project was not due solely to the use of laser sintering but also because a 'concurrent' approach was adopted throughout. The designers and casting companies all worked together to ensure that the project succeeded.

**Fig. 8.5 A range of cores produced during the Niagara Project
Courtesy VOAC Hydraulics Division**

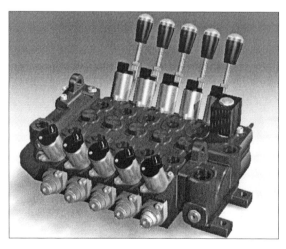

**Fig. 8.6 One of the Niagara hydraulic control valves from the Niagara
Range. Courtesy VOAC Hydraulics Division**

The chief project Engineer, Anders Karlsson, involved with the project, has since stated:

"The use of solid modelling techniques, laser sintering and resin patterns has resulted in a dramatic saving in time for production of prototype castings, with reductions from 3–6 months down to 2–4 weeks depending on the complexity of the casting

The cost for producing prototype castings with traditional methods compared to rapid prototyping has a break-even point of somewhere around 100 castings. This has given us a lot of room for design changes and early castings for verifying the hydraulic functions without major investments in pattern equipment.

Accuracy of the castings has been very good with no scrap. Some problems have been encountered with porosity, and this seems to be due to the high resin content of the sand.

Due to dramatic changes in design for the first castings we have saved over £80 000 (sterling) in pattern making costs.

Being able to produce castings so early in the project gave the production engineers an opportunity to optimize NC-paths and cutting data. They were able to suggest design changes in order to minimize future problems such as deburring and cleaning. This would not have been possible using traditional methods.

Using this technique has allowed us to delay the manufacture of production pattern equipment until designs have been verified.

One amazing benefit of this technique, which was totally unexpected, occurred with the initial samples from the serial equipment. None of the casting dimensions was outside tolerance. It is usual for the inspection report to run to a few pages on out of specification dimensions.

The use of solid modelling improves the quality of the patterns, but the use of rapid prototyping in the foundry has resolved pre-production problems with the serial pattern equipment."

The development of this partnership has continued with work on further reducing the time scales. For small volumes, the production of a complete mould and core is feasible. The data for the last core and casting manufactured for this project was provided for this trial. The data was prepared and two copes, drags, and cores produced on one build. The length of time required for building and post-processing the moulds and cores was two working days.

**Fig. 8.7 The moulds produced at the end of the Niagara Project.
Courtesy VOAC Hydraulics Division**

**Fig. 8.8 The castings produced at the end of the Niagara Project.
Courtesy VOAC Hydraulics Division**

This could be reduced down to one working day if a shift system were employed in this area of The RP&T Centres activities. Casting at a local foundry took a further two days. There is potential to produce a few castings in a week or less.

Conclusions

The use of laser sintering within the casting industry is currently in its infancy. The equipment is very costly to purchase and run. For those who are prepared to take the risk of making such a large capital investment the rewards are great.

The technique has only found use in the development of new castings and very low volume production, there is still a requirement for serial pattern equipment. Those that offer laser sintering as a service are ideally placed to produce both prototype castings and production equipment.

It should be remembered that the process of laser sintering resin-bonded sand has limitations. It is not unreasonable to suggest that in time these will be resolved and much faster machines with many banks of lasers or some other method of bonding the sand are developed. This will enable the user to produce much larger volumes or much larger cores and moulds economically.

At present there are a number of barriers for widespread application of this process, including the requirement for CAD data and the high capital cost of the laser sintering equipment. When these problems are resolved, the foundry and pattern making industry will be able to enter a new era in the manufacture of moulds.

References

(1) **Johnson, O.** and **Cant, T.** The Practical Application of Time Compression Technologies to Fast Prototyping of Automotive Metal Components, NPL Technologies Limited, Nuneaton, UK.
(2) **Wilkening, C.** Fast Production of Technical Prototypes Using Direct Laser Sintering Of Foundry Sand, EOS Gmbh and Hayward, D. HK technologies, EOS Technical Paper.
(3) **Wendt, F.** Sand Casting Moulds – Four Years Experience with EOSINT S, ACTech Gmbh, Urapid 99 London, UK.
(4) **Audisio, M.** An Industrial Application Of DCP, Erreci Rapid Casting, EOS International User Meeting 1999.
(5) **Dolner, G., Hansch, D., Trumper, J.,** and **Wendt, F.** Rapid Cylinder Heads at VW, Prototyping Technology International – issue 2.

Acknowledgements

The author wishes to thank Parker Hannifin/VOAC Hydraulics Division, Bevan Simpson Foundries, and NPL Technologies for permission to publish the Chapter and for the assistance of Mike O'Conner of Metalcast Limited and Peter Harrison of WCM Patterns Limited.

C Ryall, G J Gibbons, and R Hansell
The RP&T Centre, Warwick Manufacturing Group, University of Warwick

9

Direct Shell Production Casting (DSPC)

Y Uziel

Abstract

Direct shell production casting (DSPC) is a patternless rapid casting process, which makes conventional casting obsolete. In DSPC, the actual ceramic casting moulds, with integral cores, are produced directly from a CAD file of the cast part, without the need for patterns and core boxes. DSPC is the most flexible casting process; there is no need for draft, core prints, or parting lines. The entire gating system is designed on CAD as a part of the DSPC mould. It enables optimal feed of the molten metal directly into the critical sections of the cast part, and optimal chill placement. A complete, built-in venting system of the cores eliminates any possible out-gassing. DSPC moulds are poured hot or cold, in gravity, low pressure, or vacuum.

This Chapter describes the DSPC process, and compares it to different conventional and RP techniques for metal casting. A discussion of materials, alloys, tolerances, and applications is included.

9.1   History of DSPC

Soligen developed DSPC in 1992. Soligen was the first licensee of three-dimensional printing (3DP), a technology invented and patented by the Massachusetts Institute of Technology (MIT). The first machine (DSPC 1) was developed in collaboration with a consortium of industry partners that included; United Technologies' Pratt & Whitney, Johnson and Johnson Orthopedics, Sandia National Labs, and Ashland Chemicals. The co-development partners were interested in building a machine that would eliminate the wax patterns in investment casting. Their main concern was for the machine to develop highly accurate casting moulds with an extreme surface finish.

The initial results of DSPC cast parts exhibited very accurate and repeatable castings. Dimensionally, the ability to change scale factors made the second casting very accurate. The development team demonstrated the ability to compensate for uneven shrinkage that has always been a challenge in metal casting. However, the initial results also had some downsides. The layer building method of the shells resulted in a stair step pattern on the surface of the moulds which is replicated in the cast parts. Another difference between investment casting shells and DSPC is permeability. This required the development team to develop new methods to vent the moulds.

In 1993, at the culmination of the initial development, Soligen decided to ignore the investment casting application and focus on complex sand and lost foam casting, as well as certain permanent mould applications.

DSPC represents a possible paradigm shift in metal casting. Its effect on the end-users, as well as the foundries, could be quite revolutionary. For end-users, it is the first casting technology that enables true 'art to part' or digital manufacturing. There is no need to include draft or core prints in the design. Thus, the implementation of the CAD model, with the exception of basic castability issues such as transitional wall thickness profile, is direct and automatic. There are no 'tooling considerations' such as parting lines and loose pieces. The end-user is, thus, not required to be a casting expert.

For foundries, DSPC can augment solidification simulation and provide cast parts before they proceed on the design of the casting tools or select the casting technique. Different gating patterns and casting orientation can be tested, and result in a more optimal tooling design. DSPC can eliminate a 'soft tool' for pre-production. However, the foundry industry has yet to reach the comfort level where parts made in DSPC can be produced in conventional casting methods and result in identical performance. Such a determination could be critical in performance of cylinder heads, manifolds, or certain aerospace parts.

Realizing that paradigm shifts may take a long time with many obstacles, Soligen elected to focus on the two markets that could mostly benefit from digital manufacturing of cast parts. These are; powertrain components including water cooled or highly cored parts such as cylinder heads, engine blocks, and intake manifolds; and aerospace highly cored parts that are produced in relatively small quantities which could economically justify digital manufacturing and complete elimination of patterns or other conventional casting practices.

In 1996, Soligen produced the first multi-cylinder head with double overhead cam geometry. It was a lost foam design with very intricate water jacket and cast in oil gallery. These parts were for a demonstration of DSPC abilities. They came in pressure tight, with core cavities which could not be cast except in lost foam or, of course, DSPC. It would take five years and several pushrod cylinder head programs before the industry took advantage of the DSPC capability in developing a new lost foam DOC cylinder head.

9.2   How DSPC works

DSPC produces the actual ceramic moulds for metal castings directly from three-dimensional CAD designs. No tooling or patterns are required. DSPC is based on MIT's patented three-dimensional printing technology to produce the ceramic casting moulds for metal casting using a layer-by-layer printing process.

The first step in DSPC is to design a virtual pattern for net shape casting, including a gating system through which molten metal will flow. In addition, virtual chills are added and a core venting system, when required, is designed to optimize the solidification and eliminate castability defects (see Fig. 9.1).

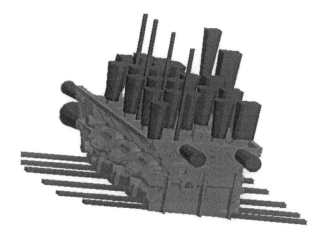

Fig. 9.1 CAD file of the gated cast part

This virtual pattern is then converted into a digital model of the mould (a CAD file of the actual ceramic mould). The CAD file of the mould, which includes integral cores, is then transferred to the DSPC system (see Fig. 9.2). The DSPC machine is a three dimensional printer which automatically generates the casting mould by successively printing cross sections of the mould.

Fig. 9.2 The DSPC system

After lowering the previous layer, a new thin layer of ceramic powder is spread. Then, liquid binder is 'printed' on to the powder layer to define a cross section of the mould utilizing a high throughput multi-jet printer (see Fig. 9.3 and Fig. 9.4).

Fig. 9.3 Multi-jet printhead

Fig. 9.4 A cross section of the casting mould is printed. The cross sections are printed layer by layer to complete the mould

This process is repeated until the entire mould is printed. The printed mould is then fired, resulting in a rigid ceramic mould surrounded by unbound powder. The unbound powder is removed from the mould by using compressed air and, for hidden sections, a vacuum cleaner (see Fig. 9.5).

Fig. 9.5 The unbound powder is removed by vacuum

Large moulds are made from multiple sections. Once the sections are clear of ceramic powder, the entire mould is assembled from the sections which register to each other with registration pins. The mould is then ready to fill with molten metal at the foundry. After the metal solidifies, the ceramic is broken, the cores are removed, and gating metal is removed to yield a finished part.

Fig. 9.6 Large moulds are assembled from sections

Figure 9.7 shows a demo DSPC ceramic mould for a cylinder head. The openings in Fig. 9.7 are for gates and risers. The water jackets and part cores seen in Fig. 9.8 are integral parts of the moulds.

Fig. 9.7 A closed demo DSPC mould

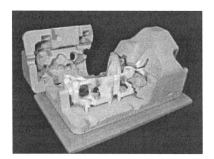

Fig. 9.8 An opened Demo DSPC mould

A typical turn around time for a DSPC made part is 12–15 business days. The CAD design of the entire DSPC mould typically takes between four hours for simple parts, and up to a week for a cylinder head, an engine block, or a complex manifold. Extreme aerospace castings could extend the design process to ten days. The printing process takes typically one day. Large parts are simultaneously printed on multiple machines. Firing the mould is an overnight process, cleaning and assembly can take another one to two days. Casting, break out and core removal, gating removal, and heat treatment requires two to four days.

9.2.1   Specifications for DSPC made parts
9.2.1.1   Materials
The ceramic powder for DSPC can be alumina or zirconium with typical particle size of 20–30 microns. The binder is colloidal silica. Chemically, DSPC moulds are similar to investment casting shells. The main difference between investment casting and DSPC is that DSPC uses a single particle size, whereas investment casting uses coarse particles in forming the external (stucco) layers. This results in different permeability. Thus, in designing DSPC, different consideration for venting must be taken.

Unlike investment casting shells which require preheating, DSPC moulds can be poured cold or hot. This results in more flexibility in making parts from different alloys and with better control of mechanical properties. (DSPC allows rapid solidification in cold moulds rather than slow solidification in typical investment casting moulds.)

DSPC moulds are more dense and do not include any organic binders. As a result, they exhibit far less out-gassing than sand casting moulds do. The alumina is not reactive with most alloys (except for titanium which requires zirconia as a mould material).

As for the casting material, DSPC has exhibited good results with all aluminum alloys (206, 319, 328, 355, 356, 357, as well as 380 and 390), magnesium, including rare earth enhanced alloys, copper brass, bronze, iron (ductile austempered, malleable, and gray), steel, stainless steels (3 xx, 4 xx, 17–4 ph, 15–5 ph), and super alloys such as inconel and titanium.

Since DSPC enables very accurate 'stitching' of sections, casting size is not a limitation. The printing envelope of the current DSPC machines is 14 inch × 18 inch × 14 inch. However, since ceramic moulds can be assembled from multiple ceramic pieces, the final casting mould can be much larger. In fact, by building the casting moulds in sections, the process provides the ability to compensate for accumulated printing tolerances. The largest casting made with DSPC was a Tomahawk missile midbody which was 2.5 feet in diameter and 7 feet tall. For automotive applications, V-12 blocks and heads were cast successfully.

9.2.1.2 Accuracy
Because DSPC uses either gravity, vacuum, or low-pressure casting processes to produce DSPC parts, normal casting accuracy rules of thumb should be observed. The DSPC process is capable of producing a net shape casting. It tends to produce moulds that are more accurate than standard sand casting. However, different geometry's may shrink differently and some sections restrain shrinkage more than others. Where special accuracy is required, these deviations can be corrected by casting a part, measuring its dimensions and modifying the scale factors of the mould. However, for tolerances on non-heat treated DSPC cast parts, refer to the following table.

Dimension	Tolerance (inches)
Equal or smaller than 1 inch	±0.021 inch
Between 1 inch and 12 inch	0.021 + 0.002 per inch
Greater than 1 inches	0.021 + 0.005 per inch
Angles	±30°

9.2.1.3 Surface finish
The default DSPC parts surface finish is 'as printed'. As these DSPC moulds are cast, some of the surfaces of DSPC parts contain 'stair steps' of approximately 0.005–0.007 inch. Shot blasting and gating removal eliminates most of the stair steps from external surfaces and near areas of gating. When required, hand finishing of DSPC moulds or the castings can improve surface finish to approximately 125–150 micro-inches rms. One DSPC unique ability is to use proprietary ceramic coatings or even high temperature glazing materials that are compatible with metal casting. This can improve surface finish of the part, including internal cavities, with more uniformity and less geometrical deviation than hand finishing or surface treatment

of the cast parts. Figure 9.9 shows a cast part with improved surface finish via glazing the DSPC mould.

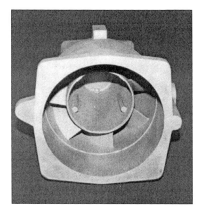 

Fig. 9.9 Improved surface finish is accomplished by glazing the DSPC mould

9.2.1.4   *Drafting blending and fillets*
DSPC castings do not require drafting and fillets, because there is no tooling *per se*, which must draw. The DSPC part will, therefore, be identical to the CAD file in this respect, i.e. drafts and fillets will be in the part only if they are in the part file. Sand castings are often modified from the original geometry so that the tooling can be withdrawn from the sand without damaging the imprint.

DSPC is no different than other casting methods in the consideration of metal flow in the mould. To avoid casting defects that result from sharp corners, the CAD file should include fillets and avoid sharp corners at the intersection of walls. Another important factor in casting design is to keep the wall thickness as uniform as is practically possible. That is because a transition from a thick to a thin wall section is sensitive to shrinkage due to uneven solidification. Sharp corners and unblended transitions are the most sensitive. To increase the likelihood of getting a sound casting, the design should call for smooth blending and as large a fillet radius as possible.

9.2.1.5   *Vacuum casting*
DSPC shells are suitable for vacuum casting which enables high definition, thinner walls, and a higher aspect ratio than those achieved in gravity casting. Since DSPC yields a solid ceramic mould with no organic binders, it does not out-gas. It also allows pouring at elevated shell temperatures (similar to investment casting).

9.2.2   Casting aerospace grades
Designing the gating in CAD enables DSPC moulds to incorporate feeding structures in the cast part in more directions and entries than conventional casting. Embedding metal chills is an additional feature that DSPC incorporates freely. These enable better control of the

direction and rate of solidification of the part as well as eliminating shrinkage and other casting defects.

9.2.3   Using DSPC for digital manufacturing

Certain casting applications require only a small number of parts with an extremely complex geometry or very high casting grade. For certain military, space, or racing vehicles, the entire production series is small and spread over time. In such cases, conventional casting requires an enormous investment in casting tools, core assembly fixtures, and producing a single sand or investment casting mould can require weeks of labour. Furthermore, once the casting tools and fixtures are made, it is often very difficult to make design changes to the geometry of the part. For these applications, DSPC is the way to economically transit into digital manufacturing where no tools are required and parts can be made to order.

DSPC produces a part that conforms exactly to the CAD file. It eliminates the lead-time and the high cost of tooling (fabrication and maintenance) and allows any number of modifications or design iterations to be implemented in real time. DSPC made castings exhibit increased accuracy, consistency, and repeatability, since the parts are made directly from the CAD files. DSPC also allows zero, or even negative draft, no parting lines or core prints. Core assemblies are also eliminated since with DSPC, cores are made as a single piece which is an integral part of the casting mould. DSPC also enables making the cores hollow which is essential in venting cores of complex aerospace castings. For parts with complex core cavities, DSPC allows core geometry's that otherwise may not be feasible.

Figure 9.10 presents an example of digital manufacturing of racing vehicles. With an extremely high rate of change in the engine configuration digital manufacturing enables improved engine performance from race to race and avoids the time and the cost of casting tools.

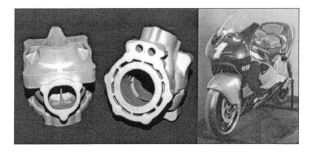

Fig. 9.10 Digital manufacturing of racing motorcycle engine (courtesy of Buckley Systems, Limited, a New Zealand Grand Prix motorcycle team)

9.3   Prototyping lost foam castings with DSPC

Recent developments in using lost foam castings make this casting technique very appealing for mass manufacturing of engine components and especially of highly cored parts such as cylinder heads and engine blocks. Using lost foam enables casting net shape heads and blocks, which result in lighter engines. The ability to eliminate conventional core prints, much of the draft and parting lines, lead to new engine designs with cast in oil gallery and much more challenging water jacket geometry. Figure 9.11 shows GM's L850, a cylinder head produced in lost foam for several years. Lost foam is an extremely automated process not suitable for short runs. Thus, making a number of prototype engines with the same geometry (especially the core cavities) is a major challenge. DSPC has been proven fully capable of making these lost foam designs for cylinder heads and engine cases for automotive and marine applications. Functional testing of engines initially produced in DSPC and later in lost foam production, have yielded identical results. This gave designers the option to complete their entire test program in parallel designing the foam, the foam tools, and the production line for those cylinder heads and blocks.

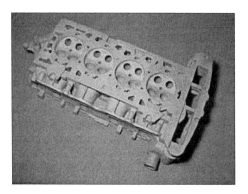

Fig. 9.11 L850 cylinder head produced by DSPC 1998 (courtesy of General Motors Corporation)

9.3.1   Using DSPC for creating hybrid castings

High-performance, aluminum cylinder heads are often cast in low-pressure permanent mould casting, or semi-permanent mould casting technique because they achieve high metal density at the fire deck (mainly in the combustion chambers). Permanent mould is made of iron or steel, both of which provide much faster chilling, which results in faster solidification and, therefore, higher density of the fire deck. Permanent moulds are expensive and time consuming. In order to overcome those limitations, Soligen has developed a technique that combines permanent mould casting, DSPC, and conventional sand casting to yield a hybrid solution. This technique allows casting short runs of cylinder heads while making modifications to the ports and water jackets. Figure 9.12 shows a racing cylinder head for a GM large block V-8. The DSPC concepts for casting these heads and achieving high density metal at the fire deck starts with casting a chill plate out of steel.

Fig. 9.12 Multi cylinder heads made in hybrid casting

Figure 9.13 shows the chill plates that were cast with DSPC and will later be used as a part of the casting tool.

Fig. 9.13 DSPC made chill plates

Figure 9.14 shows the bottom part of the casting mould. It is made with a base of dry sand that includes the chill plates in it. The chill plates are sprayed with a zirconium-based wash and include thermal heaters to bring their temperature to the desired level. The cores (for both the ports and the water jackets) are made by DSPC, thus allowing modification in CAD and the testing of different ports and water cooling geometry concurrently.

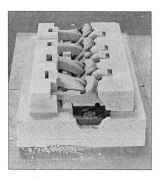

Fig. 9.14 Hybrid tool bottom half

Figure 9.15 shows the cope made in conventional dry sand as used in semi-permanent mould casting. Finally, assembly of all of these elements produces the casting mould for the cylinder heads (Fig. 9.16). The hybrid approach cuts the time and maintains flexibility in the design process. Metallurgical tests showed grain structure and mechanical properties in these hybrid cast parts as similar or superior to production cylinder heads.

Fig. 9.15 The cope

Fig. 9.16 The hybrid mould ready to be poured

Summary and conclusions

DSPC produces, directly from a CAD file, the actual ceramic casting moulds complete with accurate integral cores. By combining DSPC, conventional casting, and CNC practices, lead times, product development costs, and production tooling costs are substantially reduced. With the process, the following advantages are realized.

- *Speed* – substantial reduction of time-to-market due to the elimination of the need for tooling in the design phase.
- *Ease of making design iterations* – elimination of engineering and extensive CAD work in blending core geometry, draft angle, and core print design.
- *Cost reduction* – the ability to incorporate design changes without the cost and lead-time associated with producing or modifying patterns or tooling.
- *Design flexibility and versatility* – the ability to make complex parts with integral cores just as easily and accurately as simple parts, and to test an unlimited number of design iterations including testing a design with different alloys.
- *Seamless transition from CAD to conventional casting* – design and fabrication of production tools can be made from the same CAD file as the approved part.

Y Uziel
Soligen Inc., Northridge, California, USA

Index